FREEWAYTOPIA

How Freeways Shaped Los Angeles

by Paul Haddad

Foreword by Patt Morrison

SANTA
MONICA
PRESS

Published by: Santa Monica Press LLC
P.O. Box 850
Solana Beach, CA 92075
1-800-784-9553
www.santamonicapress.com
books@santamonicapress.com

Printed in the United States

Santa Monica Press books are available at special quantity discounts when
purchased in bulk by corporations, organizations, or groups. Please call our
Special Sales department at 1-800-784-9553.

ISBN-13 978-1-59580-101-2

Publisher's Cataloging-in-Publication data

Names: Haddad, Paul, author. | Morrison, Patt, 1957-, foreword author.
Title: Freewaytopia : how freeways shaped Los Angeles / by Paul Haddad;
foreword by Patt Morrison.
Description: Includes bibliographical references. | Solana Beach, CA: Santa
Monica Press, 2021.
Identifiers: ISBN: 978-1-59580-101-2 (paperback) | 978-1-59580-786-1
(ebook)
Subjects: LCSH Los Angeles (Calif.)--Description and travel. | Los Angeles
(Calif.)--History. | Los Angeles (Calif.)--Buildings, structures, etc. | Roads
--California--Los Angeles County. | Roads--California--Los Angeles
(Calif.) | BISAC TRANSPORTATION / Automotive / History | HISTORY
/ United States / State & Local / West (AK, CA, CO, HI, ID, MT, NV, UT,
WY) | TECHNOLOGY & ENGINEERING / Civil / Highway & Traffic |
TECHNOLOGY & ENGINEERING / Civil / Transportation | SOCIAL
SCIENCE / Popular Culture
Classification: LCC TE25.L55 H33 2021 | DDC 625.71--dc23

Cover and interior design and production by Future Studio
Maps on pages 14, 28, 54, 90, 120, 152, 166, 178, 206, 236, 258, 294, 312, 328,
by Bryan Duddles.
Front cover photo from author's personal collection.
Back cover photo courtesy of Wikimedia Commons (Bamsb900, CC BY-SA
4.0).

To my family,
for whom freeways were a transport
to some of our greatest adventures
(except the 405, which almost led to our dissolution).

TABLE OF CONTENTS

Each first date refers to the original opening segment of a freeway;
the second date refers to a freeway's completion.

The iconic Four Level Interchange, 1959

"Freeways are what we are."

—GEORGE TAKEI, "Sulu" of *Star Trek*

VALLEY DISTRICT

Pacif

Visual pres

THE LOS ANG

CENTRAL DISTRICT

OCEAN

n of

S REGION

HARBOR DISTRICT

PUBLISHED BY COURTESY OF
PREPARED IN THE
DEPARTMENT OF CITY PLANNING
by Henry V Woll

Los Angeles in 1946 before the completion of a single freeway.
Note the proliferation of oil wells in the coastal plain and agricultural fields in the Valley.

FOREWORD

by Patt Morrison

It's Los Angeles, not Rome. Here, all roads lead to... more roads. Our lore and stories speak of the beach, of Hollywood, and of that other constant on our landscape—freeways.

You know more about freeways than you think you do—and much less than Paul Haddad does. With verve and authority, he loops you in in *Freewaytopia,* an encyclopedic, anecdotal and photographic history of Los Angeles County's two dozen freeways.

Now, "-topia" can be preceded by "u-" and by "dys-," and Haddad acknowledges both. Here is homage to the lyrical prose and praise for freeways from California native and onetime Corvette Stingray driver Joan Didion, and from the influential English architectural critic Reyner Banham, who found a kind of Zen coherence in the "Autopia" of freeways.

Human lives were once described as nasty, brutish and short; our freeways, with their carnage and road rage, can be nasty, brutish and endless, and yet they have forged, for good and ill, the Southern California lifestyle in a way that even Hollywood cannot match. Our expectations for the freeways' nonstop door-to-door mobility delivers us to our destinations but also disconnects us from the happenstance of seeing and knowing the neighborhoods between points A and B—from knowing our own city.

Open to almost any page and you'll find a saga, a drama, a nugget to amaze your friends, like the many names and nicknames any freeway has had, as if they were Tolstoy heroines. The Long Beach Freeway was originally the Los Angeles River Freeway, the San Diego Freeway was the Sepulveda Freeway. (I was quoted a few times in these pages, once about my campaign for the stubby Marina Freeway to be named not for that manmade feature but for a natural one, the Ballona Freeway.)

Freewaytopia is an L.A. scrapbook of the momentous and the trivial—pranksters, mayhem, catastrophes natural and manmade, the fathers of the SigAlert and of the Botts' Dots that go bump in the night and the day, Mayor Sam Yorty's up-in-the-air upstaging of the governor of California at the opening of the 405, and why, to the perplexity of newcomers and visitors, we use the definite article "the" for our freeways, like the 210, the 405, the Harbor, the Santa Monica, the Golden State (which gave its name to a serial killer).

What elbows through these pages most profoundly and consistently, from the planning for the 1940 Arroyo Seco Parkway to the Century Freeway, opened in 1993 and correctly named the Glenn Anderson Freeway ("the Glenn," said no one, ever), is the arrogance and ruthlessness of the freeway planners and their backers.

Just like the hammer to which everything is a nail, to these builders, virtually every traffic problem could be solved by building another freeway. In the way that Robert Moses hammered apart New York neighborhoods with expressways and urban makeovers, freeway planners' policies prioritized "practicality, rapidity, and efficiency over quality of life concerns."

Part of the headlong rush to web L.A. with freeways was because a federal interstate budget at one point footed 90% of the bill for freeways, an irresistible incentive. And part of it smacked of "urban renewal" as freeways stomped through established neighborhoods like Godzilla with a cement mixer. Thousands of homes were leveled, thousands of families displaced. The Santa Monica Freeway blithely sundered Black neighborhoods near Downtown and near the ocean. The Hollywood Freeway smashed through dozens of grand old houses, two of which had belonged to Rudolph Valentino and Charlie Chaplin. The Century Freeway took out the Beach Boys' family home and shattered Black communities. Farther north, one man who had just been forced to move to make way for the Hollywood Freeway found out that he'd have to abandon his new house to make way for the Golden State Freeway.

The massive East L.A. Interchange blasted apart the character and closeness of neighborhoods, and Haddad found that real estate appraisal agents for the state's highway division had contemptuously characterized East L.A. as "infiltrated by minority groups, mostly [of] Latin derivation."

It was characteristic of the casual racism of planning that marks L.A. to this day. And when opposition to freeways reached a critical mass, with the voices of whiter and more prosperous communities, along with a burgeoning environmental and social justice awareness, freeway-building finally began to take its foot off the accelerator.

Every chapter jogged some freeway memory for me. Our spirit guides through the freeways were the air traffic reporters, with their hovering big-picture views. My friend, the pioneering KNBC anchorwoman Kelly Lange, began her broadcast career as Dawn O'Day, delivering airborne traffic reports every nine minutes in a stretchy silver jumpsuit and boots. Her headset squawked with LAPD and CHP reports and voices from her own newsroom and other stations' aerial reporters. In her four years on that job, she told me, "we had three terrifying near misses in the chopper, because all of us would cluster aloft around the biggest disasters down below!"

Their voices became as familiar to us as our own, but they took chances to tell us about the freeways that we risked ourselves on. Ex-military pilot Max Schumacher, "Captain Max," and four others were killed in 1966 when Schumacher's helicopter and an LAPD helicopter also monitoring traffic smashed into each other near Dodger Stadium.

In 1986, Bruce Wayne, KFI's "eye in the sky," crashed his plane at the Fullerton airport. His wife, Lois, went to the airport and broadcast an account of the crash and of her husband's career for KFI.

And in 1977, Francis Gary Powers, the CIA pilot shot down by the Soviets in a secret U-2 spy plane and later bartered for a KGB colonel, was flying his KNBC helicopter back from getting pictures above a brush fire when he crashed at the Sepulveda Dam. He and his cameraman were killed.

Thanks to *Freewaytopia*, I've solved the mystery of the spot on the Four Level Interchange that smells of rotten eggs—or farts. It's an ineradicable old sulfur spring, and *Los Angeles Times* photographers of yore liked to drive through that freeway tangle with a new young reporter in the front seat. They'd sniff at the smell dramatically and turn to look at the kid and arch an eyebrow, and each of us would stammer and protest—it wasn't me!

And it made me recall that when the last bit of the Glendale Freeway to the Ventura Freeway was about to open, I was the intern sent to cover it, and the Caltrans engineer let me drive, thrillingly and briefly, at 80 miles an hour the wrong way on the not-yet-opened freeway.

And now that you know them better, get out there and hit those freeways—before they hit you.

BEGIN FREEWAY

Freeways.

Few words trigger such an immediate association with Los Angeles. Despite accounting for only 527 miles of roadway in L.A.—roughly one mile for every square city mile—freeways represent a  disproportionate influence on the city's lore and populace. And yet the stories behind these universally recognized emblems are largely untold. They seem to exist merely as fixed parts of the local landscape, like the Santa Monica Mountains, the marine layer, or taco trucks—just always *there*, taken for granted and inviting little scrutiny.

As a native Angeleno who came of age in the 1970s and early '80s, freeways loomed large in my imagination. I was obsessed with their speed, efficiency, and design. I scrutinized Thomas Guides and Auto Club Metro Freeway Maps the way other kids pored over the backs of baseball cards. Once I got my driver's license, my unconditional adulation evolved into the kind of love-hate relationship most commuters have with freeways. As I reached adulthood—and began to write books and essays about Los Angeles—my fascination grew more complicated. I began to appraise them from other angles. How did they get here? Why are they where they are? And what can they tell us about ourselves? This last question strikes at the heart of why freeways matter. To know Los Angeles, one must know its freeways. They are the tableaus upon which so much of L.A.'s foundation is etched, living monuments reflecting the evolution of a metropolis, for better or worse.

Author Joan Didion is perhaps the most famous literary figure to write about L.A.'s freeway system, calling it "the only secular communion Los Angeles has" in her 1979 book of essays, *The White Album*.

It's no coincidence that the vast majority of L.A.'s freeways were built between 1940 and 1980, a period of dynamic growth that saw the city's population double from 1.5 million to 3 million. Their very form—futuristic, majestic, progressive—speaks to the seductive veneer of Southern California while engendering Eighth Wonder of the World superlatives. Having famously captured Marilyn Monroe when she was still Norma Jean Dougherty, photographer Richard C. Miller found a similar magnetism in freeways. "I just went out of my mind," he wrote upon laying eyes on the world's first four-level interchange, located in Downtown Los Angeles. "I thought, 'My God, this is what people must have felt when they first saw the cathedrals in Europe.'" Englishman Reyner Banham, author of *Los Angeles: The Architecture of Four Ecologies*, saw in our freeways "a special way of being alive," arousing "a state of heightened awareness that some locals find mystical."

Or infuriating. Freeway aestheticism is hardly top of mind for the typical L.A. driver, who spends about two workweeks a year stuck in soul-sucking traffic. Nationally, two of the five worst congestion points are on L.A. freeways—the 5 and the 101—a stat that doesn't even indict the busiest freeway of them

FREEWAYTOPIA

These young folks on the Santa Monica Freeway in 1980 have the right idea. Who *hasn't* wanted to just pull over and soak up some sun when stuck in mid-day traffic?

all, the 405. Still, those idling in the fast lane during rush hour are the lucky ones. From the 1930s through the 1980s, hundreds of thousands of residents found themselves in the unlucky position of living in a freeway's future pathway. Many were minority groups whose neighborhoods were targeted by local and federal officials as "unsightly," "unsanitary," or "blighted." After their houses were moved or razed, they were displaced for Angelenos' privilege of complaining about snarling traffic. This book will examine both the sociological and functional components— and their often uneasy coexistence—that inform the city's freeways.

While there are 24 freeways in Los Angeles County, these pages spotlight those twelve that have a sizable presence within L.A. city limits. (Apologies in advance to devotees of the Antelope Valley,

OFF-RAMP

One man built his name on traffic. In 1955, Loyd C. Sigmon of Sherman Oaks invented the SigAlert. Originally envisioned to warn residents of a hydrogen bomb attack, Sigmon's system allowed emergency personnel to alert the public of traffic snafus by funneling their reports to radio stations. Traffic jockeys would then warn listeners of, say, "a SigAlert on the Ventura Freeway near Cahuenga, both lanes." Though still in use, the term is gradually falling out of favor for other catchy names.

Pomona, Santa Ana, and San Bernardino Freeways, which reside entirely or largely outside the city.) As we dig into this Dirty Dozen, the following are a few rules of the road to aid in your reading experience. Freeway jargon can get complex, but my aim is to demystify and simplify.

Like players sporting athletic jerseys, freeways are interchangeably identified by names and numbers. The modern L.A. driver refers to them by their route designations, but several iconic ones—the San Diego, the Santa Monica, the Ventura—are also invoked by their names. For decades, freeways were typically known *only* by their names; indeed, some older directional signs omit route numbers entirely. In that spirit, we'll adopt the regional parlance and refer to freeways both ways.

There are also categories of freeways. To extend the sports analogy, imagine the "players" representing different teams, each bearing its own logo. Among L.A.'s freeways, half are interstates, recognizable by their classic red, white, and blue roadside shields. I-10 and I-5 are just two examples of national routes extending through Los Angeles.

A second group are state

This directional sign near Western Avenue and Victory Boulevard in Glendale is a living relic. It's one of the few still standing that refer to a freeway strictly by its name—sans route number.

Shields marking the three types of L.A. freeways, left to right: Interstates, State Routes, and U.S. Highways.

highways. Unlike interstates, state highways don't *have* to be freeways (think State Route 1, otherwise known as Pacific Coast Highway). Or, as in the case of SR-2, they can start as a street (Santa Monica Boulevard) and turn into a freeway (Glendale Freeway). A handful of L.A. freeways are state routes, whose green, spade-shaped insignia mimic the shovels of "forty-niner" gold miners.

The third type are U.S. Highways, or U.S. Routes. While these traverse multiple states (like interstates), they sometimes feature cross-traffic (like state routes). The 101 Freeway is a U.S. Route, its digits framed by a three-pronged, crowned crest.

While it's useful to distinguish these differences, I promise no pop quizzes on route nomenclature. Like the old *SNL* skit, *The Californians*, freeway numbers are tossed around with casual disregard to their classifications.

As part of our journey through L.A.'s freewayopolis, this book will also explore routes that never saw the light of day. Planners initially envisioned 1,500 miles of freeways in greater Los Angeles—three times the number actually built. Imagine superhighways burrowing through Laurel Canyon, tunneling under the Hollywood Sign, or spanning the waters of Santa Monica Bay. With a few more legislative strokes of the pen, you wouldn't have to imagine them; they'd already exist. A cousin to this "what if" club are zombie freeways, or "ghost" freeways—a term

OFF-RAMP ↗

It's an age-old question: Why do Southern Californians put a "the" in front of freeway route numbers—as in, "Take the 5 Freeway to the 101"? The answer lies in the early practice of assigning freeways names to counter their multiple route numbers. It was less confusing to say "the Golden State Freeway" instead of "the 4-slash-6-slash-99 Freeway," reflecting the patchwork of routes it went by. During the Great Renumbering of 1964, routes were consolidated under one numerical designation, making it easier to use their numbers—as in, "the 5 Freeway," which eventually was shortened to simply "the 5." The "the" simply stuck around from when they were all called by their full names.

OFF-RAMP ↗

The longest freeway that originates in Los Angeles County is the San Diego Freeway, running 150 miles and ending in San Diego. Believe it or not, the 2.5-mile Marina Freeway is not the shortest. Three freeways are so short, they can be covered in a brisk walk (I would not recommend this): the Terminal Island Freeway (1.6 miles), the Colorado Freeway (the 0.6-mile Colorado Street exit off the 5), and the Metropolitan Bypass Freeway (a 1.5-mile exit off I-5 in Gorman, part of the unbuilt High Desert Corridor).

popularized by former Caltrans engineer Arturo Salazar—whose vestiges exist as stubs or ramps that either never joined living freeways or were put out to pasture. They too get their due alongside our marquee freeways.

I'm particularly interested in giving space to lost voices who bring every dimension of the freeway experience to life: Minority neighbors who rallied against eminent domain. Women engineers who thrived in a man's field. Elected officials who helped further freeways, stop them, or broker happy mediums between their constituents and the government. And the corps of civic and state highway employees whose collective vision, skill, and hard work produced not just the most famous freeway network in the world, but feats

A 2020 study estimated that 6,000 to 7,000 homeless people live along freeways. For decades, freeway bridges have provided shelter and, in some cases, electricity as some encampments access the roadways' power sources.

A fan of photo ops, California Governor Pat Brown often flew down from Sacramento to christen freeways with leggy beauty queens bearing cumbersome titles. Here he cuts the ribbon on a new segment of the Santa Monica Freeway in 1962 with, among others, Miss Harbor Freeway, Miss San Diego Freeway, and Miss East L.A. Interchange Freeway.

of engineering that, at their best, achieve architectural poetry. So integrated are their concrete creations in the local landscape, each has taken on a personality of its own, creating a kind of emotional validation loop between motorist and motorway.

Speaking of which, it can take years—sometimes decades—for freeways to be completed. For clarity, the chapters of this book are laid out in a chronological timeline based on each freeway's opening segment, rather than its completion date; these "sneak peeks" to the public were often as short as one mile. (The Century Freeway was the only freeway to roll out in its entirety.) The dates "1938–1953" in the Arroyo Seco Parkway chapter heading, for example, indicate that its first stretch opened in 1938; the entire route wasn't drivable until 1953. Stringing our narrative through these initial segments allows us to better track the tandem development of Los Angeles and freeways, like a historical documentary unspooling through the flickering lamp of a 16mm projector. Our imaginary screen will reveal all the fanfare that accompanied these freeway ribbon-cuttings—including enough beauty queens to fill an old Miss America pageant.

OFF-RAMP ↗

Since 1972, Caltrans is the recognized shorthand for the California Department of Transportation, which oversees the state's highways (freeways are a form of highway). Previously, the agency was called the California Division of Highways under the Department of Public Works. The name change was to underscore the state's updated multi-modal approach to transportation.

Starting with the Arroyo Seco Parkway's Rose Queen, no freeway could open without them. We'll also see how attitudes toward freeways shifted from largely idolatrous to increasingly acrimonious.

But first, the central question: Where did freeways come from?

Like many, I used to believe their origin story sat squarely with Germany's autobahn system. There's some truth to that, but like a lot of things . . . it's complicated. Contrary to legend, the Reichsautobahn was not invented by Adolf Hitler, having predated him by a few years. But when Hitler assumed power of the Third Reich in 1933, he became its biggest champion. In 1935, a newly commissioned fourteen-mile autobahn between Frankfurt and Darmstadt impressed the world. The *St. Louis Post-Dispatch* rhapsodized about this new type of automobile mega-highway. "It Will Only Be Used By Motorists (and the Military)" read the paper's headline.

The parenthetical was telling. By December 1941, Germany's 2,400 miles of proto-freeways were almost *exclusively* used by the military, enabling swift ground transport against contiguous European countries. Ironically, this efficient system factored into the Nazis' downfall. When General Dwight Eisenhower's forces invaded Germany, his trucks found pristine strips of concrete to mobilize across the countryside—far superior to the rutted, bombed-out roads of war-torn France. After the Allied powers vanquished the Reich and occupied Germany in 1945, Eisenhower never forgot this battery of uninterrupted thoroughfares.

Meanwhile, during Europe's war years, scenic parkways were already operating in the New York area under roadway mastermind Robert Moses. But with maximum speeds of thirty-five miles per hour and only two lanes in each direction, these were really just "embryonic freeways," as Caltrans engineer Heinz Heckeroth termed them. Here on the West Coast, the Arroyo Seco Parkway, dedicated in 1940, looked more like the traditional freeways we see today. Whereas autobahns and Moses's parkways were largely rural, the Arroyo Seco Parkway linked two urban areas while still retaining a curvy, Sunday Drive sensibility with a speed limit of forty-five miles per hour.

As cities across America were choking on congestion, what was *really* needed were conveyances that could accommodate high speeds (like autobahns) within urban districts (like the Arroyo Seco Parkway). On March 30, 1946, the Los Angeles Metropolitan Parkway Engineering Committee called out this transit void when, responding to a fact-finding mission by California Senator Randolph Collier, it presented a study called *Interregional, Regional, Metropolitan Parkways in the Los Angeles Metropolitan Area*. The group concluded that "the trend of transit . . . is for mixed traffic expressways or parkways not only in the more remote sections of the area, but predominantly so right through our cities and towns."

As the chairman of the state's Senate Transportation Committee, Collier spearheaded an action plan. In 1947, the "Father of California Freeways" partnered with Assemblyman Michael Burns to create the Collier-Burns Act, paving the

No politician did more to build California's freeway system than State Senator Randolph Collier, right, huddling here with Assemblyman Tom Carrell, left, and Governor Pat Brown.

President Dwight Eisenhower cruises Los Angeles alongside Governor Goodwin Knight in September 1954. Within two years, Ike would sign into law his signature act that created the nation's interstate system.

way for thousands of freeways financed through gasoline and vehicle taxes. With Sacramento now calling the shots, the Department of Public Works conceived a galaxy of freeways in the state's urban areas, including Los Angeles.

Six years later, Dwight Eisenhower began his first term as the thirty-fourth president of the United States. Though there was already legislation for a nationwide highway system, the Eisenhower Administration made its passage a priority. Using California as a model, lawmakers ditched the idea of tolls while the general-turned-president flashed back to World War II. "After seeing the autobahns of modern Germany and knowing the asset those highways were to the Germans, I decided, as president, to put an emphasis on this kind of road-building," Eisenhower said. "Germany had made me see the wisdom of broader ribbons across the land."

The result was the Federal Aid Highway Act, also known as the National Interstate and Defense Highways Act. Signed into law on June 29, 1956, it allocated $25 billion for 41,000 miles of interstate across the nation—the largest public works project in American history. Combined with the Collier-Burns Act, this was another shot in the arm to Southern California freeway-building. For proof, take a gander at any "Proposed Freeways" map from 1958. You'll find the Southland awash with black route lines. In the plans, ten freeways/expressways blast through the Santa Monica Mountains alone (thankfully, only two survived the planning stage: the San Diego Freeway and the Hollywood Freeway). The Interstate Act dangled another incentive: the federal government would pay, minimally, 90 percent of the costs. With California on the hook for only 10 percent of the bill, a mad scramble ensued to build interstate freeways before the highway

This municipal map from February 28, 1958 outlines at least 1,500 miles of freeways in L.A. County. The thick dark lines represent freeways already adopted or budgeted by the state. Gray and broken lines are proposals for future freeways. Only one-third of these routes were ever built.

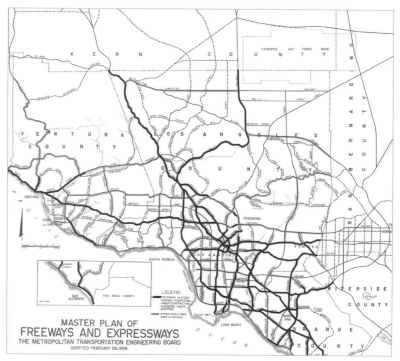

OFF-RAMP ↗

As with autobahns, Eisenhower foresaw a potential military application for interstates (accounting for the "Defense" language in the act). "In case of an atomic attack on our key cities," said the president, "the road net must permit quick evacuation of target areas [and] mobilization of defense forces. The present system would be a breeder of a deadly congestion within hours of an attack." There you have it—traffic can be fatal.

trust ran dry . . . or Uncle Sam changed his mind.

It was another president, John F. Kennedy, who popularized the phrase "victory has a hundred fathers, and defeat is an orphan." During the heady years of freeway adoptions—the 1950s through the mid-1960s—the Los Angeles Freeway System was considered the pinnacle of success. Famed 1960s art historian Kurt von Meier saw them as "architectural events . . . with the inventive and creative qualities of works of art." Eisenhower, Collier, Moses, and a consortium of German engineers all contributed to its advancement, but so did endorsements by local chambers of commerce, business groups, council members, the voting public, and the Southern California Automobile Club, which published its own freeway blueprint way back on April 16, 1937, and

L.A.'s freeways are not for everyone, as tourist Charlie Abboud discovered in 1959. According to the *Los Angeles Times*, the Nebraska native spent "eight terrifying hours as a captive on the freeway system," after which an LAPD lieutenant tried to straighten him out. Postcards from the 1960s and '70s played up the glut of "crazy Los Angeles freeways."

Next stop, Freewaytopia.

provided early roadside signs. "The climate in Southern California was one of accepting the vision of freeways," recalled Heckeroth, who had a hand in every freeway from the Arroyo Seco to the Century and designed the East Los Angeles Interchange—the busiest freeway nexus in the world.

Embryonic freeways were built by many hands. But like the Cobb salad, gangsta rap, the modern skateboard, and the Hollywood studio system, the modern freeway was born in Los Angeles. With that, its exquisite grand dame is ready for her close-up . . .

Chapter 1

THE ARROYO SECO PARKWAY
1938–1953
State Route 110

Imagine stepping into a time machine that whisks you back to early 1940s Los Angeles. You climb into your Cadillac custom convertible—top-down, of course—and you and your squeeze rumble out of your Eastside bungalow for a Sunday drive. The sky's a smog-free cobalt blue, palm trees are lolling in the breeze, and the radio is thrumming with Tommy Dorsey. Chocolate phosphates would be good about now. You hop on Figueroa for the twenty-minute drive over to Fair Oaks Pharmacy in South Pasadena.

But what's this? It's that new "stopless motorway" you've heard about! How did they make it stopless?

You give it a whirl. It's beautiful. Lots of Beaux-Arts bridges and straightaways mixed with gentle curves. The wind nips your face as your speedometer touches new heights. *What a rush!* You pull up outside the soda fountain in eight minutes. You've seen the future. You vow never to drive on surface streets again.

It's appropriate that Los Angeles's love affair with freeways began with the loveliest of them all—the Arroyo Seco Parkway (aka the Pasadena Freeway, as it was known from 1954 to 2010). But you weren't the only one who experienced their first freeway crush. So did a million and a half other Angelenos. And like any love affair that endures, familiarity breeds contempt. Sure, you changed, but so did your beloved freeway system. Traffic got between you. Finally, after decades of growing old together, you vow never to drive on a freeway again.

L.A. motorists' conflicted relationship with freeways is

rooted in their dependency. It's simply hard to get around Los Angeles without them. But as a living relic that predates the nuclear age, L.A.'s oldest freeway gets a pass. Hating the Arroyo Seco Parkway is like hating your grandma. She is of a different epoch. Angelenos don't just accept her anachronisms; they've come to treasure them. In the last few decades, the Arroyo Seco has nabbed as many accolades as Katharine Hepburn did Oscars, recognized as a California Historic Parkway, National Civil Engineering Landmark, National Scenic Byway, and National Register of Historic Places. All this, and somehow the freeway only seems to get better with age.

For that, we can hate her.

The first proposed vehicular corridor from Pasadena to Downtown wasn't a highway, parkway, or motorway. In fact, it had nothing to do with cars. It was a bikeway.

Fifty years before the Arroyo Seco Parkway opened, the modern-day bicycle was invented. These "safety bicycles" had air-filled tube tires and equal-sized wheels, which was far more practical than those penny-farthings with colossal front wheels you see in old pictures (usually with intrepid mustachioed men atop them). Men, women, and even children took to these new bikes, launching the Great Bicycle Craze of the 1890s.

Looking to rake in on this rage sweeping the country, an intrepid, mustachioed businessman named Horace Dobbins envisioned an elevated bicycle route along the Arroyo Seco riverbed. Horace secured six miles of right-of-way for his California Cycleway, running from the Green Hotel (now Castle Green) in Pasadena to Avenue 54 in Highland Park. Alas, only one and a quarter miles of his wooden cycleway ever got built. Pasadenians soured on the ten-cent toll just as the bike fad was fading. Though his venture failed, Dobbins's right-of-way did establish a route for

Horace Dobbins's California Cycleway around 1900, looking north toward Pasadena. Bicyclists appreciated its smooth surface and unobstructed elevated lane; however, many were put off by its tolls and having to lug heavy bikes up fifty feet of stairs.

the future Arroyo Seco Parkway.

As the cycleway was being dismantled in the early 1900s, Germany opened the world's first mass-transit suspended railway in Wuppertal, still in use today. This early monorail inspired stateside inventors, who claimed to have the answers to L.A.'s increasingly clogged streets. In 1907, Joseph W. Fawkes pitched a sixty-miles-per-hour monorail running from Burbank to Los Angeles. A few years later, Fletcher J. Felts tried to sell investors

Inventor Joseph Fawkes's experimental monorail—dubbed the Aerial Swallow—whisked along passengers from 1910 to 1912. But it never got farther than his own private orchard, and subsequently collected rust after Pacific Electric extended a railroad line to Burbank.

on a Pasadena-to-Downtown aerial line that could hold up to 100 passengers per pod. These contraptions never stood a chance, but clearly Fawkes and Felts were onto something. For the next hundred years, monorails would continue to intrigue us as a primary or supplemental mode of transportation alongside freeways. Expect them to pop up several more times in this book.

Meanwhile, Horace Dobbins took one more crack at an Arroyo Seco route. Back in his day, the Santa Fe Railroad took up to forty-five minutes to wend through the canyon (a trestle from 1896, now bearing MTA's light rail line, still spans the freeway). Dobbins's newly formed Pasadena Rapid Transit Company imagined an express train that could make the nine-mile trip from Pasadena to L.A. in twelve minutes—a direct path along his rights-of-way through a series of elevated tracks, tunnels, and open cuts into the hillside. But Pasadena citizens voted down a bond that would have authorized its construction in 1919. Once again, Dobbins had whiffed. If he had been paying closer attention, he'd have seen that trends were changing.

At the dawn of the Roaring Twenties, trains and streetcars were *so* 1910s. The County of Los Angeles had added 400,000 people in each of the previous two decades, reaching just under 1 million by 1920. As the city's tentacles spread outward, automobiles became essential, whisking Angelenos to places where creaky trolleys couldn't. Early on, of course, Henry Huntington's Pacific Electric Red Cars contributed to L.A.'s sprawl. From Sierra Madre to Redondo Beach, the transportation and real estate mogul lured prospective homeowners by shrewdly laying down tracks to his newly created subdivisions—only to become a "victim" of his own success. Thanks to an oil boom and newfound affordability, car ownership skyrocketed. By the mid-1920s, there were 200,000 cars a day swarming Downtown, leading to one of L.A.'s most enduring features: the traffic jam. Meanwhile, Pasadena tallied 27,500 registered cars, one for every 2.4 citizens— the highest per-capita ownership rate in the *world*. Trolleys

Northeast view of the Arroyo Seco before construction of the parkway. Note the foundations in the lower-middle, where a Pacific Electric railway bridge was removed to clear a path. A small sign near the flood control channel reads: "PWA Federal Works Agency Public Works Administration."

went from quaint to annoying to perilous, impeding the free flow of autos. Something had to give.

Enter the brain trust of Frederick Law Olmsted Jr., Harland Bartholomew, and Charles H. Cheney. In 1924, the city planners mapped out a vision for Los Angeles in their seventy-page *Major Traffic Street Plan*. Included in their study was a proposal for a parkway through the Arroyo Seco, which would "create the feeling of openness that comes only with plenty of width and by an ample enframement of trees, shrubs, and other plantations." This wasn't the first time someone had suggested a scenic motorway from Pasadena to Los Angeles along the "bed of the stream." At the turn of the century, both the city of Pasadena and L.A.'s Municipal Art Commission fully subscribed to the City Beautiful movement, an ideal rooted in inspiring urban projects to promote mental and physical well-being. On November 4, 1924, voters approved legislation to build a parkway along the arroyo.

A MODEL FOR THE FUTURE

But what exactly *was* a parkway? That depended on who was doing the talking. Olmsted and company defined it as a functional high-speed road that elicited "a great deal of incidental creation and pleasure." In this way, it was not all that different from the parkways being built in New York—divided-lane, limited-access superhighways with an emphasis on pleasure-driving.

OFF-RAMP ↗

Other picturesque roads were also proposed in the early 1900s. The journal *California Outlook* pushed for a series of byways linking most of L.A.'s lake parks. One plan called for a Silver Lake Parkway connecting Silver Lake with Echo Park Lake, Westlake (now MacArthur) Park, Elysian Park, and Griffith Park. In 1914, Los Angeles mayor Henry Rose nixed the parkways after right-of-way acquisitions proved problematic.

Gradually, the term became interchangeable with "freeway." Contrary to popular belief, a freeway is not a roadway free of tolls. The term is used to describe any restricted-access highway in which abutting property owners have no rights to impede the free movement of cars passing through (I know, dull, right?). Over time, freeway planners started to emphasize speed over ornamentation and mental welfare. Of course, the Arroyo Seco would have all of the above—thus, the swappable terms—although newspapers and mapmakers would continue to apply the "parkway" designation to other nascent freeways, regardless of their aesthetics.

Financing for this parkway/freeway—or "stopless motorway" as some still awkwardly called it—was established in 1923 when the state tacked on its first gas tax. (Initially, this tax was two cents a gallon. That's right, the next time you curse California's high gas prices, you can blame the Motor Vehicle Fuel License Tax Act.) Washington also kicked in capital. During the Great Depression, the Works Progress Administration—part of Franklin Delano Roosevelt's New Deal program—bankrolled the Figueroa Street (future freeway) viaduct over San Fernando Road and the

On the left, construction of the southernmost Figueroa Street Tunnel, circa 1936. The ornate tunnels were converted to the Arroyo Seco's northbound lanes when the freeway opened in late 1943.

Los Angeles River. And just as the Army Corps of Engineers cemented L.A.'s municipal waterways in the '30s, federal agencies pumped $7 million into shoring up the Arroyo Seco. Despite its name, which translates to "dry stream," the seasonal channel was prone to deadly flash floods. Without the government taming its banks, an Arroyo Seco Parkway would not be viable.

Fortunately for head engineer Lyall Pardee and his team, the route's proximity to the riverbed skirted dense neighborhoods. But in what would become a familiar pattern, progress meant uprooting some unfortunate souls. L.A. commissioners simply handed over 380 acres of city land near the Los Angeles River—the initial southern terminus—which included 125 buildings. Because they were not located in the more desirable woodsy area of the arroyo, officials felt they were not of considerable value.

Farther south, between the L.A. River and Sunset Boulevard, engineers scoped out a future expansion of the freeway through the well-established Elysian Park. Eighty-nine lots were condemned, affecting dozens of apartment units, some housing extended families of fifteen to twenty tenants. They were given twenty days to move after their buildings were acquired. If they didn't like it, well, too bad; under new legislation, the state granted the Division of Highways the authority to purchase and demolish properties along a freeway's path. And so began a pattern that would continue for decades where, as Pastor Stephen "Cue" Jn-Marie of Los Angeles's Row Church related, "the linkages between highway construction and the removal of Black and brown folks was a frequent theme for those who stood to profit in state and federal road-building programs."

Those whose homes were spared did not necessarily breathe a sigh of relief. Pasadena and South Pasadena were among the wealthiest enclaves in the nation, with lots of old-money families skeptical of projects for the public good, especially if the WPA was involved. They circulated petitions opposing the parkway, citing its inevitable dead-end streets, air and noise pollution, and general disruption caused by new construction. Mostly they worried about depreciating property values from motorists bypassing local businesses. Officials assured them that the opposite would happen. Based on similarly constructed motorways on the East Coast, citizens should expect *increased* land values and business tax revenues, as well as improved access to their neighborhoods. But would gains in revenue offset losses in property taxes from razed homes? Variations of this debate between locals and highway officials would flare up with each new project over the next half-century.

Nonetheless, enthusiasm for the Arroyo Seco Parkway ran high. Despite a nationwide depression, it represented the unlimited promise of Los Angeles, a chance to step out of the long

Two ancient bridges spanning the parkway are the 750-foot long steel Santa Fe trestle (now a Metro line), built in 1896, and the York Avenue arch bridge, which was originally a wooden trolley viaduct. It was converted to concrete in 1912.

OFF-RAMP ↗

Before WWII, when a knock on your door by the military meant a death notification, few things struck more fear than a visit from a right-of-way agent. Frank Balfour was one of the state's first, forming a union and becoming somewhat of a legend. His beat was Southern California, where, swooping down from Sacramento, he was referred to as "the wind from the north." He retired as chief of the Right-of-Way Department in 1960, having blown through neighborhoods abutting every L.A. metro freeway up to that time.

shadow of its northern rival, San Francisco (which had completed the Golden Gate Bridge in 1933), with a signature engineering triumph of its own. Angelenos had already glimpsed the future with the Ramona Freeway, itself a sneak-peek of the Arroyo Seco Parkway.

Opened in 1935, the Ramona stretched four miles east of Downtown along today's I-10, transforming Ramona Boulevard into a roadway with all the hallmarks of a freeway: separate grades, sloped embankments, and bridges replacing cross-traffic. But the Ramona traversed an industrial strip. The Arroyo Seco would add beauty to the mix and be built from scratch. "There isn't anything like it in the country!" enthused Justus Craemer, an engineer from the Department of Public Works.

As the parkway neared completion, Craemer invited members of the press on drive-alongs to pump up the public. One such passenger was the *Los Angeles Times'* Ed Ainsworth, who hosted a column called "Along El Camino Real." In an entry from March 29, 1936, Ainsworth turned Wordsworth, waxing poetic about what he saw. "The road

OFF-RAMP ↗

Freeway crusaders were equally vocal, distributing leaflets entitled "FREEWAY TRUTH" that extolled their economic benefits. These groups pushed for a northern extension that would eventually become the Foothill Freeway (I-210), but their proposal was roundly rejected by voters in 1937. By the way, the Arroyo Seco Parkway will never reach the 210 Freeway, because it does not meet federal interstate standards. This is why the 710 (which is an interstate) was always the preferred route to close the infamous freeway gap to the 210.

scoots up the channel like a scared jackrabbit . . . a beautiful drive between oaks and sycamores most of the way," he said. "A marvel of ingenuity . . . the model for the future." Lanes and ramps were banked to counteract the centrifugal forces of fast-moving vehicles. Ainsworth particularly liked the planting of shrubs along the center median to shield oncoming headlights, and the lack of inter-

Signage for the Ramona Freeway at a junction southeast of Downtown Los Angeles. Two miles opened to traffic in 1935, emanating eastbound from Aliso Street. Its route tracked portions of US 99 (now I-5) and US 60 and US 70 (now I-10), which became the San Bernardino Freeway. At that point, the nearby Pomona Freeway claimed the "60" designation as a State Route.

sections that "eliminate all the monkey business of stop signs, turns, and traffic jams." It was a no-brainer to place more parkways throughout the region. "If they don't grab the wings this super-road provides they ought to go back to riding in a surrey."

Starting in December 1938, two years before the official dedication, officials began to open small sections of the Arroyo Seco to motorists. As a real-time petri dish, engineers could now see what was working . . . and what wasn't. The good news: traffic volume was higher than expected. The bad news: the parkway, built to safely accommodate 27,000 cars per day, was already outdated. To help the situation, shoulders were converted into travel lanes. (Of course, this eliminated shoulders entirely . . . a void that continues to this day.) And those beautifully manicured shrubs along the median? They quickly became weeds, growing into the roadway.

The lanes were another quirk. To (subconsciously?) discourage drivers from straying into other lanes, engineers used two

Early lanes of the parkway resembled an Oreo cookie—a white inner lane sandwiched by two dark ones—in an attempt to dissuade lane-switching. A far bigger problem was the lack of a center guardrail in exchange for ornamental shrubs—a custom that lasted well into the 1960s on some freeways.

different shades of concrete, alternating the colors. Thus, each three-lane direction had black concrete lanes on the left and right and a white concrete lane down the middle. Lane coloring was never proven to reduce accidents, and the practice was nixed for future freeway surfaces. Even the Arroyo Seco went all gray in short order.

LET THERE BE TRAFFIC

"It takes courage to do a thing for the first time." The proclamation was made by Governor Culbert Olson on December 30, 1940, during the official coronation of the Arroyo Seco Parkway. The "first" was the first *urban* limited-access, grade-separated roadway in the United States—an honest-to-goodness freeway, minus the annoying tolls of Moses's parkways.

In some clever marketing, the freeway's debut was moved up to capitalize on all the media already in town for Pasadena's annual Rose Parade and Rose Bowl. Los Angeles was no stranger to splashy openings, of course. One month earlier, two dozen klieg lights lit up the night sky around the Carthay Circle Theatre for the red-carpet premiere of *The Great Dictator*, Charlie Chaplin's much-anticipated first talkie. But excitement for L.A.'s revolutionary roadway was even greater. It directly affected people's lives. Officials from Los Angeles and Pasadena felt it their duty to put on a show that met the moment.

The day began with a ceremony at Los Angeles City Hall, from which a caravan of VIPs motored to the freeway's entrance at the Figueroa Street viaduct. Clearly there were some first-day jitters. Cruising along at the speed limit of forty-five miles per hour, the procession somehow managed to get into three fender benders. Eventually, they made it to the freeway's end at Glenarm Street in Pasadena, where 1,500 spectators gave them a hero's welcome. Joined by county, state, and federal politicos, Governor Olson kicked things off by introducing 1941's Queen of the Tournament of Roses—Sally Stanton, a seventeen-year-old, apple-cheeked darling wrapped in a ring of roses and a fur-collared coat.

Though she didn't realize it at the time, Stanton was a pioneer. No dedication over the following decades was worth its weight unless it had a beauty queen (or several of them) flanked

Throngs of the freeway-curious take in the dedication of the Arroyo Seco Parkway on December 30, 1940. Governor Culbert L. Olson praised not only its utilitarian value, but also its beautiful features that "delight the eye of the artist."

Everything's coming up roses for Rose Queen Sally Stanton and her assemblage at the ribbon-cutting. They include, two men to her right, Highway Commissioner Amerigo Bozzani and, to her immediate left, Governor Culbert L. Olson, as well as assorted highway patrolmen.

by a contingent of older white male officials to christen a new freeway. At one point, a policeman admonished an unknown man for touching the Queen. Turns out, he was her father.

The pageantry rolled out like an elaborately staged play. Balloons, flag-raisings, military bands, and color guards struck up a celebratory tone. The Goodyear Blimp circled the sky, capturing images for the state's promotional film, *California Highways*. And then, a turn toward solemnity. Chief Tahachwee, in full American Indian headdress garb, blessed the freeway on behalf of his Kawei forefathers who used to roam the land. This had become routine for the chief. During the soft opening in 1938, he had transferred his tribe's interest to the state's highway department with a ritual that included a drum circle and the sharing of peace pipes. Cringeworthy as it seems now, these showy acts of inclusiveness played into audiences' romanticized views of simple savages in an ever-changing world while also assuaging white guilt.

Properly sanctified, the freeway was now ready for its ribbon-cutting. Sally Stanton was granted the honors. In a photograph blazoned across newspapers, she struggles to snip the red

OFF-RAMP ↗

Southern Californians have cycled through various names for the state's native peoples. The "Kawei" lexicon for tribe members in the Pasadena and South Pasadena regions was phased out long ago. Today they are recognized as a group of Tongva Native Americans known locally as the Hahamongna—also the name of the former settlement-turned-parkland near the confluence of the 210 and 134 Freeways.

silk ribbon with giant scissors. After Governor Olson pulled on the ribbon to make it tauter, the Rose Queen finally succeeded. Olson then waved at highway patrol chief E. Raymond Cato, who led the first civilian motorists down the southbound ramp onto the black-and-white stripes of the Arroyo Seco Parkway—a six-mile journey that could be made, as one publication marveled, in the time it takes to smoke a cigarette.

The entire project came in at the bargain-basement price of $5,050,000. Only 10 percent went toward rights-of-way acquisitions, an extremely low percentage compared to future freeways as the Southland got more built up. The governor watched the traffic disappear toward the City of Angels as if seeing a kid off to college. "This is, to say the least, most extraordinary," he said. "The dream has been a long time dreaming."

The parkway's debut got surprisingly scant attention outside of California; it was, at best, a sidebar to the bigger story of Stanford's win over Nebraska in the Rose Bowl. But it did impress C. E. McBride of the *Kansas City Star*, who correctly predicted

Sally Stanton and Governor Olson. Stanton would go on to participate in various Arroyo Seco ribbon-cuttings over the next fifty years.

In a familiar charade, Chief Tahachwee shares a peace pipe with Director of Public Works Frank W. Clark at the December 30, 1940 ceremony. Off-camera, the beating of tribal drums signified the formal transfer of property rights from the Kawei Indians to the state of California.

that the freeway "is more than likely to be copied in one way or another by all our great cities as the ever-increasing traffic problems come up for solution." He particularly admired the freeway's remove from urban degeneration: "Not a store, not a hotdog stand, not a gaudy billboard along the way." Indeed, the Arroyo Seco is consistently cited as the genesis for future freeways in America and a model for urban interstates.

Engineers had little time to savor their triumph. The egress of screaming southbound traffic onto surface streets resulted in multiple pile-ups at the intersection of Figueroa Street and Riverside Drive. The solution was a 2.2-mile southerly extension to Downtown Los Angeles. America's involvement in World War II actually hastened this stretch. Due to the roadway's potential to "rapidly transport soldiers and equipment," the War

Jutting off to the right, the two-way North Figueroa Street Bridge would be integrated into the 110's northbound lanes; the lanes curving left became the ramp to I-5. Before the parkway was extended southward, this chaotic intersection with Riverside Drive—now bridged by the southbound lanes— was among the city's most dangerous.

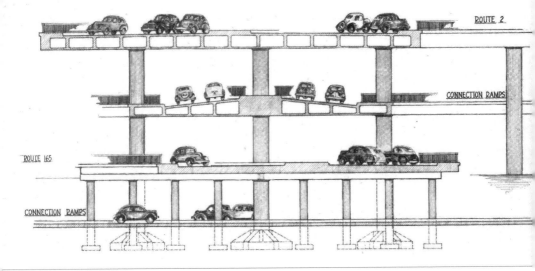

An early cross-section sketch of the Four Level commissioned by the Department of Public Works. It rendered the Hollywood-Santa Ana Freeway on the top level and the Arroyo Seco-Harbor Freeway on the third level. (The route numbers are their old state designations.) The other two levels contain transition ramps.

Department arranged for immediate federal financing. On the downside, other freeway plans throughout the region were indefinitely mothballed as labor and raw materials were funneled into the war effort.

Fortuitously, gas taxes and L.A.'s treasury paid for the rest of the Arroyo Seco's $4 million spur. Its four Art Deco-flavored tunnels through Elysian Park were already built in the 1930s. Initially a conduit for two-way traffic on Figueroa, they were converted to northbound freeway lanes. Engineers considered tunnels for the southbound lanes as well, but opted for an open-cut approach through the hillside instead, saving $1 million. A second bridge over the L.A. River and a grade separation with Riverside Drive were also built.

On December 30, 1943—once again, just before New Year's to accommodate the crush of cars to Pasadena—the southern leg of the parkway opened to the public. It stopped near College Street, about a half-mile north of Downtown's future Four Level Interchange. There was a different "feel" about this segment of the Arroyo Seco. According to the Historic American Engineering

Bird's eye view of the Arroyo Seco Parkway cutting through Elysian Park, including two of the four tunnels for northbound traffic. The exposed southbound lanes are on the left.

Record, "the urgency of the extension's construction and the dearth of 'beautification' associated with it relative to the initial development . . . helped transform this part of the Arroyo Seco 'Parkway' into something that would later more closely resemble the freeways of the Los Angeles metropolitan freeway system." It represented, in essence, the passing of the parkway, both in practice and in parlance. In a few short years, the original Arroyo Seco Parkway would come to resemble, as *L.A. Times* columnist Patt Morrison later mused, "a museum piece, dated as a zoot suit, narrow, awkward, pokey."

The contrast became even starker upon the inauguration of the so-called Four Level in 1949. Recommended by engineer W. H. Irish, it was the first interchange in the world with four stacked roadways (detailed more in the Hollywood Freeway chapter). It would

OFF-RAMP ↗

The California Highway Commission—referenced a lot in this book—is now called the California Transportation Commission. Either way, commission members are appointed by the governor and oversee the funding and general management of California's highway system. Think of them as the boss of the engineers who work at the Division of Highway (and later, Caltrans).

not attain full operating capacity until September 22, 1953, when the Arroyo Seco Freeway (as it was alternately called) finally plugged into its web of thirty-two lanes. Accordingly, this date marked the Arroyo Seco's official completion. Final length: 8.2 miles. More importantly, it now provided a seamless transition to the Harbor Freeway through Downtown, as well as ramps to the Hollywood and Santa Ana Freeways.

For the sake of clarity, one year later the California Highway Commission changed the names of several finished and unfinished freeways to reflect their destinations. The Ramona became the San Bernardino, the Sepulveda became the San Diego, and the Los Angeles River (Freeway) became the Long Beach. This is also when the Arroyo Seco Parkway became the Pasadena Freeway—a moniker it would carry for over half a century. The whitewashing of the past was now complete.

Of course, tapping into L.A.'s expansive freeway grid exposed the shortcomings of the Pasadena Freeway's original six-mile country ramble. To keep up with the Joneses, the freeway's max speed was raised from forty-five to fifty-five miles per hour. Lanes were widened to twelve feet. Though there were still no shoulders, fifty emergency turnouts—modest pockets every few hundred feet—were added. One peculiarity that persists to this day are the freeway's shockingly short access ramps. Endearing as they are, they were already sorely antiquated by the mid-'50s, leading to several fatal accidents. Because they couldn't be lengthened, engineers added stop signs to the on-ramps, forcing motorists to achieve freeway speeds with virtually no runway and zero momentum.

As for its plant-lined, rolled-curb median, it was toothless in stopping motorists from careening into

OFF-RAMP ↗

One thing that has fortunately survived after all these years is the "City of South Pasadena" sign, made out of stones and positioned on the south-facing berm of the Arroyo Drive Bridge. Dating to the 1930s, the sign's flat rocks were extracted from the Arroyo Seco's watershed.

opposing traffic. After several grisly wrecks, the Pasadena, like other freeways, eventually converted to real barriers.

ARTWORK ZONE AHEAD

Perhaps because it was the first freeway, the Arroyo Seco/Pasadena Freeway gets outsize attention every time it reaches a major anniversary. Each milestone offers an opportunity to measure this engineering case study against the passage of time. Its twentieth anniversary—celebrated on December 30, 1960, on the Sunset Boulevard overpass—was a strange brew of nostalgia and space-age aspirations. Former Rose Queen Sally Stanton was brought back to cut another ribbon and present a plaque. Now married with kids, she would go on to have a rewarding career as a mathematician at Jet Propulsion Laboratory, though she would always be known as the First Lady of ribbon-cuttings. She continued to grace anniversaries with her presence, including the freeway's fiftieth.

State bureaucrats used the twenty-year anniversary event to herald the road's safety and efficiency. According to Harrison R. Baker—a former member of the California Highway Commission—350 million vehicles had traveled the freeway between 1940 and 1960. Though it handled "three times the number of cars at twice the average speed" as surface streets, its accident rate was "five times as favorable." He estimated that the freeway had saved $54 million in costs related to gasoline, maintenance, time savings, and accident reduction. He also claimed that one life per year was saved for every mile of freeway built, a figure that highway officials liked to float.

The next twenty years would be even more critical. A roadway equipped to handle a daily flow of 27,000 cars was now clogged with 70,000 and rising fast. How would state officials keep the aging matron from complete obsolescence? Baker sparked a conversation that carried well into the mid-1960s, an "anything

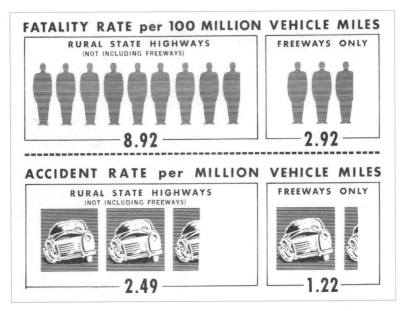

FATALITY RATE per 100 MILLION VEHICLE MILES

RURAL STATE HIGHWAYS
(NOT INCLUDING FREEWAYS)

FREEWAYS ONLY

8.92

2.92

ACCIDENT RATE per MILLION VEHICLE MILES

RURAL STATE HIGHWAYS
(NOT INCLUDING FREEWAYS)

FREEWAYS ONLY

2.49

1.22

These stats were produced by the California Highway Commission and appeared in the Jan./Feb. 1960 edition of *California Highways and Public Works*. Traffic studies from the publication were often disseminated to the public to tout the benefits and relative safety of freeways vis-à-vis surface streets.

goes" era that envisioned freeways over the ocean and as fifteen-mile tunnels. In 1965, the *Pasadena-Independent* outlined some of the more outlandish proposals, most of them centered on renting "space rights" over the Pasadena Freeway. Because it was largely a sunken route, one could imagine all sorts of uses for covering its six lanes—parkland, parking lots, housing tracts, landing pads for helicopters, even runways for small planes. Interestingly, none of these ideas to "put a lid on it" addressed the central issue—traffic—and seemed better suited to *The Jetsons*, which just may be where they came from.

Meanwhile, the mid-'60s recalibration of route numbers across California included the Pasadena Freeway. Since its opening, portions of the route went by various labels, including U.S. 66, U.S. 99, U.S. 6, and U.S. 11. In 1964, the state reassigned Route 66 (yes, *that* Route 66) to the Pasadena/South Pas portion of the freeway. It then transferred State Route 11 from Figueroa Street (now tagged with Route 66 Alternate) to the strip south of the Figueroa viaduct. Just to add to the confusion, guidebooks of that period often designated the *entire* length of the Harbor and Pasadena Freeways as SR-11, a numerical appellation that would remain fixed until 1981. When the Harbor was classified as a federal interstate, it switched to

This "ghost ramp" is the original Fair Oaks Avenue off-ramp off the southbound lanes. The ramp was virtually hidden around a blind curve and caused safety concerns; the Fair Oaks off-ramp was subsequently relocated half-a-mile north, before the curve.

I-110. To keep the numbering consistent from San Pedro to Pasadena, the Pasadena Freeway—still a state route—became SR-110. The number 110 began to appear on green freeway signs, outlined by the familiar spade symbol, and for twenty years, no one was confused about it.

Then, on the morning of August 5, 2001, a California Department of Transportation worker—decked out in a company-issued orange vest and white helmet—clambered onto the catwalk of a freeway sign. His assignment: to alleviate confusion pertaining to the northbound 110's overhead freeway sign near Downtown's 4th Street. The problem wasn't what was *on* the sign; it was what *wasn't*.

The sign directed motorists to Pasadena via the 110 Freeway, with three arrows pointing to the three left lanes. The Caltrans employee, working alone, affixed two metallic plates above the far-left arrow: an Interstate 5 shield and, above that, the word "NORTH." Driving through one day, I remember noticing the change and thinking how thoughtful it was that Caltrans finally added "5 North" to let people know they needed to take the 110 to get to the 5 North cutoff.

Fast-forward nine months, when a local newspaper published a scoop that had even Caltrans scratching their heads. The sign installer did not work for the agency and was not authorized to carry out any such job. In fact, he was an imposter. A trespasser. A guerrilla artist. As video from that day later revealed, he could be seen driving up to the freeway in a truck whose door sign cleverly advertised "Aesthetic De Construction."

The altered freeway sign that bore artist Richard Ankrom's call to "guerrilla public service." The "North 5" plates stood for nine years before Caltrans unveiled an updated sign that preserved Ankrom's contribution.

His name was Richard Ankrom. The sandy-haired, lantern-jawed artist was inspired to amend the sign based on his own disorienting experience navigating this stretch of freeway through Downtown. He saw his stunt as a public service. Like a skilled jewel thief, Ankrom had planned his sleight-of-hand for months. He carefully studied the Caltrans road signs—specs, color palettes, types of sheet metal—then manufactured and painted the signs himself. A small video crew documented his caper. Finally one of the crew members, unable to keep the secret any longer, leaked the truth to the *Downtown News*. The story went national, showing up on *Today* and *Good Morning*

OFF-RAMP ↗

Horace Dobbins never did finish his California Cycleway from Pasadena to DTLA. But in 1995—about one hundred years later—an avid bicyclist named Dennis Crowley vowed to finish it for him. After the Metropolitan Transportation Authority rejected his plan for a designated bike path, Crowley hit up private investors to come up with the estimated $10 million needed. He even got then-mayor Richard Riordan interested. The plan fizzled, but expect another push to revive the Cycleway in 2095.

The Arroyo Seco Parkway's "modern" segment, west of the 5 Freeway, facing the Downtown skyline. A classic L.A. snapshot, encompassing multiple eras of progress and the coexistence of urban and natural environments.

America, and Ankrom was hailed a freeway folk hero.

How did Caltrans respond? After inspecting Ankrom's handiwork, they deemed it up to safety standards and allowed his sign to stand, with no charges filed. What else could they do? They had been duped themselves. In fact, Ankrom's addition was validated in 2010 when, as part of routine maintenance, Caltrans replaced the sign with a more reflective one that prominently included "5 North" alongside "110 Pasadena" on the green field. As for Ankrom, he continued to commit stealthy acts of public-service art, always waiting seven years before confessing to any mischief. As he told *ABC7 News*, "I have to wait for the statute of limitations so I don't go to jail."

WHAT'S OLD IS NEW

At the start of the 2010s, everyday life was suddenly suffused with nostalgia. Recession-reeling Americans found comfort in artisanal gin, retro shows like *Mad Men*, and anything with Betty White, who popped up in Super Bowl ads and *SNL*. Vinyl records continued their improbable comeback, and *The Walking Dead*—a visceral throwback to George A. Romero's zombie movies—exploded into our living rooms.

Improbably, the Arroyo Seco Parkway also came back from the dead, part of a re-brand by Caltrans that kicked the Pasadena Freeway to the curb after fifty-six years. Preservationists rejoiced at the freeway's reversion to its original name, which really sank in when all the signs were swapped out in the summer of 2010. Our desire to reconnect with the past is often more instinctual than logical, and the 110 was no different. For most of my lifetime, the Pasadena Freeway was maligned. But the Arroyo Seco Parkway? *That* sounded like a return to romance.

If Caltrans made one mistake, it was that they simultaneously announced a $17 million overhaul to make the freeway safer. Suddenly, motorists realized they were attached to all the things that were changing. Letter-writers to the *L.A. Times* decried plans for "more attractive walls" (what was wrong with the old ugly ones?), replica lampposts (they aren't authentic enough), and concrete center dividers (bring back the steel and wood ones!). "The Arroyo Seco Parkways is a real gem," said Paul Daniel Marriott, the author of *Saving Historic Roads*. "I don't think people fully appreciate that yet and I think they'll regret the damage that's being caused." Even replacing the original slanted curb along the median drew howls of protest, though others didn't see what all the fuss was about. "The name should stay the same," a younger motorist told the *Times*. "We know freeways by their numbers. Our parents were the ones who knew them by name."

Fortunately, most Angelenos recognize the uniqueness of L.A.'s first freeway—the only freeway in the United States that fulfills the "scenic, natural, historic, cultural, archeological, and recreational qualities" of a National Scenic Byway. Despite the Arroyo Seco Parkway's years, it has weathered old age well. Engineers attribute its good bones to decades without trucks (banned for safety reasons) and the fact that it ends in a residential neighborhood, which discourages through traffic. Indeed, the average

June 18, 2003: More than 2,000 pedestrians and 3,500 cyclists take over the empty SR-110 Freeway during Arroyo Fest.

number of daily cars seems to have leveled out at around 125,000.

In 2020, plans were to eliminate traffic entirely—at least for a day. An organization called Active San Gabriel Valley got permission to shut down the freeway to vehicles to stage Arroyo Fest—a sequel to the first wildly successful festival in 2003. As with that event, thousands of people, mostly millennials, embraced the rare opportunity to pound the parkway on foot and admire the road's fanciful features up close. The COVID-19 pandemic postponed it, but as appreciation for urban landmarks grows, you can be sure it will be back.

That same year, my hard-to-impress sixteen-year-old daughter's first experience behind the wheel on the Arroyo Seco Parkway prompted unexpected glee. "It's so cute," she said. "It's like driving on a miniature freeway." Her embrace of its quirks recalled the rush of adrenaline I first felt as a newly licensed sixteen-year-old leaving games at Dodger Stadium by the former Figueroa tunnels. The whole ritual of negotiating a torrent of traffic from the Solano Avenue on-ramp—from a dead stop, between two freeway tunnels, with virtually no runway—was like blasting off an aircraft carrier. I'm not as fearless as I was then, but the thrill of the old road is always there in ways rarely found on other roadways, perhaps akin to what early motorists discovered themselves while sampling L.A.'s first freeway.

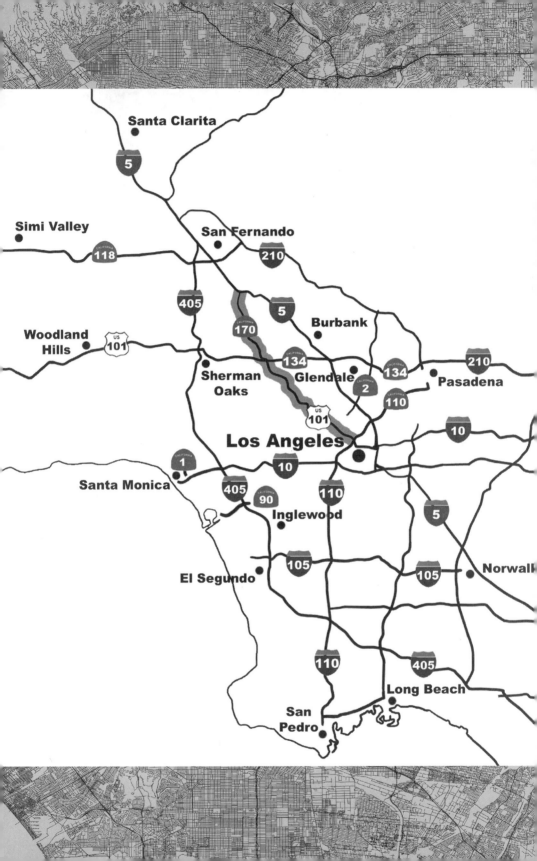

Chapter 2

THE HOLLYWOOD FREEWAY
U.S. Route 101 / State Route 170
(1940–1968)

On May 4, 1968, the *Valley Times* announced upcoming nuptials in their News and Features section. The "July wedding" would take place just before Independence Day. The 500 anticipated guests would include a who's who of the San Fernando Valley's finest, and a bevy of beauty queens would be whisked in on jet helicopters and in antique cars.

The union—surprise of surprises—was between the Hollywood Freeway and the Golden State Freeway. For the first time, Angelenos could zip continuously from Downtown to the northern regions of the San Fernando Valley.

Blame the name, perhaps, but the Hollywood Freeway was no stranger to the spotlight during its twenty-eight years of maturation. In typical fashion, this drama queen of a freeway had one more bout of drama left, one last stab at grabbing headlines before settling into a humdrum existence as just another freeway in a city now full of them. Minutes before this consummation of thoroughfares, a wedding crasher appeared overhead. His name was Larry Boberman. He was piloting a single-engine Cessna 150, out for a joyride with his eighteen-year-old girlfriend, Linda Rodewald. But he was in trouble. The plane's engine began to sputter. He needed a place to land—pronto.

Just south of the dedication site, the Hollywood Freeway provided sanctuary. Boberman spotted a clearing in traffic between Oxnard and Victory Boulevards, and glided the bird down safely. Police officers helped him push the aircraft off the freeway to a towing yard.

The next day, the San Fernando Valley woke up to the *Valley Times* article about the hitched highways. Next to the article was a headline that upstaged it in terms of sheer drama: "TEENAGE PILOT USES FREEWAY TO LAND SAFELY."

The Hollywood Freeway had delivered one final scene-stopper. Continuing a pattern going back decades, it was accompanied by the kind of glory that was ready-made for Tinseltown.

It began, as many great stories do, with a curse. For anyone who has driven through the Cahuenga Pass during rush hour, it could be argued that the curse never left.

The legend was launched in the 1860s, about a hundred years after Gaspar de Portola led the Spanish military through the pass. As with most curses, this one's origin has shifted over time. One story involves a shepherd named Diego Moreno, who doubled as a secret agent for Mexican president Benito Juarez. Moreno stole $200,000 in gold and gems from other agents who were transporting a war chest from San Francisco to Mexico to finance Juarez's resistance of French forces in his country. When Moreno reached the rutted road of Cahuenga Pass—the mountainous corridor that connects the San Fernando Valley to Hollywood—he stopped at a tavern and found a room. Whatever he imbibed must've played on his dreams, because that night, Diego had a vision of being killed if he dared set foot in Los Angeles with his

Cahuenga Pass, 1892. Those who couldn't afford to overnight at the local inn camped by the side of the road.

OFF-RAMP ↗

El Camino Real, or "the Royal Road," approximates the old route that linked the Spanish missions and presidios of Central and Southern California.

pilfered loot. The next morning, he buried it under a tree.

Continuing on to L.A., he suddenly came down with a lethal illness. Just before he died, he confessed to his friend Jesus Martinez about the buried treasure. Martinez and his stepson set out for the cache. But before Martinez could start digging, he suffered a violent seizure and he too met his maker. His stepson quickly bailed the scene, only to be gunned down in a family argument.

In another version of the fable, a Mexican servant from the San Gabriel Mission stole $300,000 worth of coins and jewels from the padres' old mill. He fled by mule up El Camino Real, which forms the spine of the 101 Freeway. When our nameless servant reached Cahuenga Pass, he was killed in a botched robbery. Legend has it he sensed trouble and stashed the booty under a tree. The assailants never found it, though they likely gained a mule.

Word of both fellows' misfortunes was passed down over generations. Whichever tale one believed, many Mexican migrants took a "pass" on the pass as recently as the early 1950s, when Pacific Electric solicited laborers to dig up trolley tracks along the Hollywood Freeway. They were worried about unearthing the spoils and disrespecting the dead.

While some feared the cursed treasure, others exploited its possibilities. In 1939, Ennis Combes—a mechanic armed with a metal detector—claimed to have pinpointed its source under the parking lot of the Hollywood Bowl. He even secured

OFF-RAMP ↗

Adding to the rumors of buried treasure, in 1885, a Basque shepherd whose flock grazed the hillside of the eventual Universal City found a weathered leather satchel filled with coins and jewels. His dog dug it up from under a tree.

permission from the County Board of Supervisors to commence digging, along with a promise to share the riches. But at the last minute, Combes got cold feet. What if he was next in line to be stricken dead? He took his magic rod and hightailed it back to Bakersfield.

Combes's business partner, Henry Jones, harbored no such fears. Later that year, he enlisted a man with an "electrochemical recorder" to finish the job. After drumming up publicity through a couple of Hollywood has-beens, Jones set up a circus-type atmosphere that drew throngs of spectators, film crews, and CBS Radio. The digging went on for twenty-four days, resulting in a forty-two-foot-deep empty hole—and Henry's abject humiliation. Shortly thereafter, he committed suicide. The curse, it seemed, was alive and well, though that hasn't stopped other treasure hunters from throwing good money at digging permits in the years since.

A CLAIM TO ROMANCE

Colorful lore aside, Cahuenga Pass is steeped in actual historical significance, particularly with respect to the Mexican-American War. It is here where Lieutenant Colonel John C. Fremont and Mexico's General Andres Pico memorialized the treaty that

This mural, painted by Hugo Ballin in 1931, recreates Mexico's surrender to the United States. Before the treaty was signed, there had been numerous skirmishes between Californians and Mexicans. Cannon balls have been uncovered in hillsides near the freeway, believed to originate as far back as 1831's Battle of Cahuenga Pass.

ceded California to the United States in 1847, setting the stage for statehood three years later. Campo de Cahuenga, the adobe ranch house where the Treaty of Cahuenga was signed, still stands in the smoggy shadow of the freeway opposite Universal Studios.

Its place in history secured, newspapers and civic boosters came to regard Cahuenga Pass as a totem of the city's romanticized past. As rumblings of a new kind of motorway began in the 1920s, media outlets poured on the nostalgia, which served as collective memory ownership for Anglo Angelenos. The fact that the pass could be traced to an old Tongva trail only strengthened the manifest destiny element. Readers were reminded of its role in the 1849 California Gold Rush: the famous frontiersman Kit Carson used to canter through the pass to deliver mail to Monterey. John Butterfield was here, too, his covered wagons overnighting at the Pass Hotel or Eight Mile House along a babbling brook. Ana Begue de Packman, matriarch of the Historical Society of Southern California, fondly recalled yodeling around blind curves to alert approaching parties, so narrow was the Old Pass Road of the 1890s. Even the war that led to Mexico's relinquishment of California was characterized as "more picturesque than sanguinary," a description that reduces the Cahuenga Treaty to a whitewashed portrait of two dapper diplomats shaking bloodless hands alongside blooming poppies and riparian chaparral.

The trajectory toward a high-speed artery was inevitable. Post-WWI, Southern California was the largest oil producer in the world. With so much cheap, plentiful fuel available, one out of every three Angelenos owned an automobile, dropping trolley ridership 75 percent from its peak. A 1924 study revealed ten times more cars in L.A. than were on the road in 1914. When the Hollywoodland subdivision opened in the Hollywood Hills, its Beachwood Drive realty office produced glossy brochures that crowed of its location high "above the traffic congestion, smoke, fog [smog], and poisonous gas fumes of the lowlands." (Such pristine conditions promoted "lusty lungs.")

When Carl Laemmle opened his early version of Universal Studios in 1915, he built grandstands for paying visitors to watch silent movies being filmed on soundstages. Audiences were urged to cheer the hero and boo the bad guy. The letters scrawled on the right read "Entrance to the City of Wonders."

It wasn't just Hollywood lowlanders who were drowning in traffic. As anyone familiar with California's water saga knows, Los Angeles had annexed the San Fernando Valley in 1915, which instantly tripled its population. The City of Angels was now bifurcated by the Santa Monica Mountains—the Valley to the north, and the original basin to the south. Suburbia arrived in the form of communities called Van Nuys, Pacoima, Encino, and Chatsworth, and others known today by different names: Girard (Woodland Hills), Lankershim (North Hollywood), Roscoe (Sun Valley), and Owensmouth (Canoga Park). Many Valleyites worked "over the hill." Their commutes contributed to Angelenos' familiarity with two of the most dreaded words in their vocabulary: rush hour.

Agriculture was the Valley's biggest business, but filmmaking was gaining steam. The same year that Los Angeles absorbed the Valley, Carl Laemmle debuted his 230-acre Universal City Studios on the north slope of the pass, in a former sheep pasture. Cecil B. DeMille was a neighbor. The legendary director rented a wooden cabin in the pass and rode his horse into his Hollywood studio, cowboying up with a revolver on his hip.

By the mid-1920s, Hollywood's myth-making machine was the fifth-largest industry in the nation, the Valley its de facto backlot. Downhill from Hollywoodland, an even swankier enclave opened up to the creative community. Whitley Heights counted Charlie Chaplin and Rudolph Valentino among its famous neighbors. Upon leaving their respective Xanadus in the mornings, each had to regularly schlep (or be schlepped) into the Valley to punch the clock, be it as the Little Tramp in *The Gold Rush* or, in Valentino's case, as a Latin lover riding his horse through the San Fernando Mission, whose grounds doubled as the Argentinian countryside in the World War I epic *The Four Horsemen of the Apocalypse*. While the Valley may have been the perfect setting for horse-and-buggy Westerns, there was nothing charming about the fact that the road out there was equally outdated—dusty, pockmarked, prone to muddy floods, and befouled with horse dung.

City officials understood Cahuenga Pass's importance to the smooth flow of goods and people and began a series of improvements. In 1922, the narrow canyon walls that once harbored Mrs. Packman's echoes were widened to one hundred feet. The pass's 745-foot apex was smoothed down from a 10 percent grade to 5 percent. But just as the Golden Gate Bridge is never finished being painted, Cahuenga Pass proved equally insatiable with each successive upgrade. By the late '20s, 26,000 cars traveled through it every day. Within ten years, volume would grow another 50 percent.

But relief was in the works. With the Arroyo Seco Parkway coming along—and the four-mile Ramona Parkway creating good buzz—city officials

Cars motoring along Cahuenga Pass before the freeway. The flat grade of Pacific Electric's interurban railway—a fixture in the pass since 1911—is visible on the far right. The tracks would later be moved to the center median of the freeway.

committed to a Cahuenga Pass Parkway that would link Hollywood with Mulholland Drive. Paid out of the state's motor vehicle tax fund, it would be just the first stage of many. Granted, its 1.5 miles was a modest goal, but in the context of the Great Depression that was ravaging the country, it was better than nothing at all.

By early 1940, the Hollywood Freeway's progress muscled its way onto local front pages alongside updates of the war brewing in Europe. The project called for four lanes of unimpeded

Lights... camera... excavate! Even before it opened, people were enamored by the star attraction known as the Hollywood Freeway. This SoCal Auto Club photo was taken on the corner of Fountain Avenue and N. St. Andrews Place.

traffic in both directions. Pacific Electric Railway tracks—present in the pass for decades—would be rerouted to the center median. The price tag? A cool $1.5 million, well beyond what could be covered by gas taxes. As with the Arroyo Seco Parkway, the Public Works Administration pitched in, closing the monetary gap from a $7 billion fund designated for civil projects between 1933 and 1944. District chief engineer Spencer V. Cortelyou sweated the details and approved the construction firm of Radich and Brown for the job. The government had just one stipulation: the Cahuenga Pass Parkway needed to be completed by June 15, 1940, or the aid would be withdrawn.

It was the perfect dramatic opening scene. Like that timeworn movie cliché, the freeway was the damsel in distress tied to railroad tracks, the feds were the oncoming freight train, and

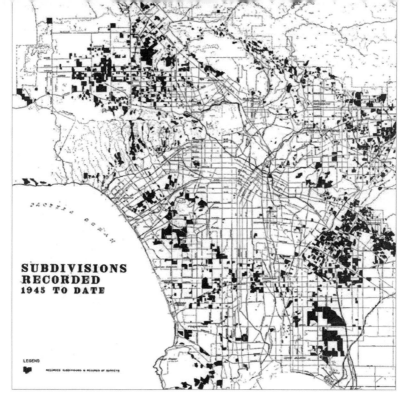

SUBDIVISIONS
RECORDED
1945 TO DATE

LEGEND

The City Planning Commission recorded new subdivisions in metro Los Angeles. The surge of suburbia is stark on this map representing 1945 to 1952. The San Fernando Valley occupies the upper left, eagerly awaiting the arrival of freeways like the Hollywood (the diagonal line in the center stopping just short of it).

the hero was Radich and Brown. The firm literally worked continuous twenty-four-hour shifts, feverishly trying to rescue the freeway before it was too late.

Not only did they finish on time, but the freeway was also dedicated six months before the rose-adorned ceremony that crowned the Arroyo Seco Parkway. The resulting spectacle set a high bar for future freeway openings, equal parts small-town parade and garish Hollywood premiere. First came the History of Transportation cortège marching down the freeway: American Indian runners, Mexican soldiers, horseback riders, Butterfield stagecoaches, and early and late-model automobiles. Every travel mode of yore was honored, though a caravan of camels representing a short-lived means through the pass in the 1850s was disappointingly absent.

The rituals then migrated to the Pilgrimage Bridge, the first

Cowboy hat in hand, Gene Autry snips a film "ribbon" to unveil the first section of the then-Cahuenga Pass Freeway on June 15, 1940. Tom Keene (and his horse) appears on the left. In the background, onlookers watch from the Pilgrimage Bridge.

overpass to span the speedway. In full cowboy regalia, B-Western star Tom Keene and his higher-wattage colleague Gene Autry—each an honorary mayor of a Valley community—held up a long, 35mm motion picture "ribbon." The actors were joined by other honorary mayors, civic boosters, Governor Culbert Olson, and state highway muckety-mucks. A phalanx of photographers and reporters captured the moment when Mayor Fletcher Bowron and Autry, the Singing Cowboy, opened their scissor blades over the film strip. Ready . . . set . . . *snip*!

The *Los Angeles Times* made the event their lead story in the Sunday morning edition, pouring on the superlatives as the pass's predecessors were, once again, resurrected:

Old Don Gaspar de Portola would have rubbed his travel-weary eyes in amazement . . .

The swashbuckling Butterfield stage drivers would have been equally startled . . .

And the pre-pueblo Cahuenga Indians would have looked for a new trail if they had all appeared anew yesterday at a familiar haunt—Cahuenga Pass.

A photograph facing south from Pilgrimage Bridge shows a

smattering of cars christening the speedway. Down the middle are Pacific Electric's twin Red Car tracks. The most interesting detail in the picture is in the upper part of the frame: the railroad passes over a tunnel, or a "road subway" as it was called then. About a couple hundred feet long, this subway funneled southbound traffic onto Cahuenga

OFF-RAMP ↗

The Hollywood Freeway was assigned the number "101" by the American Association of State Highway and Transportation Officials (AASHTO), the entity that guides highway design, construction, and numeric designations. US 101—which runs the length of the West Coast states in some form or other—is the westernmost major U.S. federal highway.

Boulevard, which swooped under the freeway while other traffic flowed onto Highland Avenue. The tunnel would last only twelve years, eliminated after the trolley was abandoned.

Meanwhile, the photo caption conveyed that the freeway was suited for 50,000 daily cars. City engineer W. E. Brown touted that it would accommodate any increases in traffic "for a long time." This was not true, unless you define one year as "a long time." As would become habit, officials underestimated the causal relationship between freeways and motorists. Freeways don't just accommodate heavy usage—they facilitate it. The *American*

The old "road subway," which whisked southbound cars under the Red Car tracks that ran up Highland to the middle of the freeway. The tunnel was narrow and not well lit, leading to many accidents, some blamed on motorists' disorientation while driving in or out of the darkness.

A Pacific Electric car traveling down the center median in October of 1952, facing north from the Mulholland Bridge. The line had a stop under the Barham Bridge, which at the time had stairs leading to the tracks.

Economic Review calls this phenomenon "The Fundamental Law of Road Congestion," whereby adding lanes or another arterial road only brings more traffic. Combined with the region's growth, traffic on the 101 increased by an average of 10,000 cars per day every year from 1940 to 1955.

Nonetheless, the freeway was an engineering feat ahead of its time: the first high-speed road in the nation to integrate rapid transit along its center median. This novel idea thrust the Hollywood Freeway into the national spotlight. General Electric—which had a vested interest in maintaining electric streetcars—produced a gauzy promotional film called *Way of the Future* that signaled dual transport as a panacea to the increasingly complicated relationship between autos and local rail. It would be another fifty-five years before the city revisited the concept, with the opening of the Green Line along the median of the Century Freeway (I-105) in 1995.

Still, it wasn't long before all the revelry produced a hangover effect. Despite being hailed a triumph of American ingenuity, the freeway was seen by some as a form of fascistic overreach. Freeways didn't bring freedom. They brought destruction, slow-moving concrete-and-steel tsunamis that wiped out anything in their inexorable path. A parkway through Cahuenga Pass was one thing—the route was largely rural, and rights-of-way were easily attained. Gouging out a 300-foot-wide swath from Hollywood to the Los Angeles Civic Center (the eventual southern terminus) would not be so simple.

OFF-RAMP ↗

For a brief spell, the state Legislature toyed with converting the freeway to a toll road. The idea of paying to access Cahuenga Pass was not unprecedented. In 1870, a toll station appeared near the Pass Hotel to help maintain the then-wagon road.

The monumental undertaking would displace scores of renters, homeowners, and businesses in some of the most historic and densely populated areas of L.A. Projected estimates placed the number of affected individuals as high as 28,690 by the late 1940s. Protestors formed an association whose grievance was right in its name: the Hollywood Anti-Parkway League. In the summer of 1940, its president, A. B. Schaeffer, filed a complaint with the County Board of Supervisors. Members of the alliance called on elected officials and potential candidates to reject the "archaic, outmoded, and un-American" appropriation of public funds to buy up property rights for unnecessary motorways.

But the freeway foes were howling in the wind. The city celebrated another ribbon-cutting in November of 1940, when the northern spur reached Vineland Avenue. Motorists could now cruise from the Hollywood Bowl to the Valley in four minutes. After the Vineland opening, construction on the Hollywood Freeway

A heart-wrenching photo of tenant Margaret Voze on November 3, 1945. After being informed that her apartment building was being cleared for the freeway, the 83-year-old clutched a cross and prayed. She lived at 226 North Olive Street. The entire north section of Olive no longer exists due to the 101.

Top photo: The freeway looking south, with the Mulholland Drive overpass in the foreground, early 1941. Bottom photo: The same shot from years earlier, when railroad tracks occupied the strip now served by the freeway.

Two views looking southeast at the Los Angeles Civic Center, the latter showing the finished freeway through Downtown in 1953.

entered a dormant stage during World War II. But this didn't stop progress altogether. Much of the early- and mid-1940s was spent acquiring rights-of-way to extend the route Downtown. As with most eminent domain cases, it was a long, ugly dogfight that stripped the engineering marvel of its alluring sheen.

The first major casualty was Whitley Heights. For years, its vocal neighborhood association had pushed highway officials to study alternate passages. One route followed the present course of Highland Avenue, and then curved southward between Hollywood and Sunset Boulevards—around the bulbous mound that was Whitley Heights—before rejoining the current path at Van Ness Avenue. But this alternative was deemed too disruptive to Hollywood's buzzing central business district. Besides, unlike the Arroyo Seco Parkway, the Hollywood Freeway was not meant to be a jaunt through the country. New policies called for freeways to seek the fastest, most direct routes from Point A to Point B.

State highway engineer A. D. Griffin tsk-tsked as his crews tore down "many beautiful homes" (ninety in all) in one of the "fine old residential sections of Hollywood." Rudolph Valentino's mansion and Charlie Chaplin's home were among those leveled. The freeway chopped the neighborhood bluntly in two, resulting in twenty-seven dead-end streets. Irate locals called the work an "outrage," a "mutilation," and "sheer vandalism." Writing to the *Valley Times* in 1948, Margel Gluck from Whitley Heights griped, "Are we living in a Police State? We just fought a war to put a stop to dictatorship." She accused the state of acting in bad faith through forced home sales. "Men were overheard to say that this would show Whitley Heights who was who, and that no protests would do any good." Gluck's suspicions were validated when two state right-of-way agents were sent to prison. Their crime: conspiring with real estate brokers to sell them the properties at jacked-up prices.

Supporters of the Hollywood Bowl—across the street from Whitley Heights—were similarly aggrieved. The approved route would carve into its parking lot. Traffic noise would permanently disrupt the famed amphitheater's finely tuned acoustics. C. E. Toberman, president of the Hollywood Bowl Association, spoke for many when he bemoaned the L.A. City Council's recent decision to scrap the Red Car line as part of the roadway's southern expansion. "A rotten engineering blunder," he fumed, citing a journalist's spurious claim that rail would "accommodate five times the number of travelers as could motor traffic alone."

So who *did* kill the Red Car? Pose that question about L.A.'s streetcar system in general and one usually gets a multitude of *Roger Rabbit*-type conspiracies. In the case of the 101, there was an urgency to hasten the freeway's progress without further cost overruns. Due to Hollywood's topography, the line would have been graded at a different level than the roadway. This would have delayed construction several years at millions of extra dollars (to support a private enterprise, no less). After granting the go-ahead without rail, the council threw rapid transit proponents a bone when they directed highway engineers to build bus turnouts to promote motor coaches. Turnouts are still part of the freeway today—for example, those "Buses Only" signs you see near the Western and Vermont off-ramps, the latter of which includes stairs down to the center median from the Vermont overpass.

After construction resumed in 1946, the Hollywood Freeway enjoyed a new burst of PR from the *L.A. Times*. On August 20, 1948, the business-friendly paper devoted an entire photo spread to its progress. The first two paragraphs of an accompanying article set the breathless tone:

> A giant furrow is being plowed through Nature's and mankind's barriers between the San Fernando Valley and the coastal plain carpeted with Los Angeles commerce and industry . . . which will carry more vehicular traffic faster than any arterial man has yet fashioned.
>
> Less than halfway completed, it looked, here, like the trail of a Kansas tornado; there, like a gigantic gob of

September 1949: The only time you'll see the Four Level Interchange without traffic. Completed later that year, the Bill Keene Memorial Interchange, as it is now known, would not service all four connecting freeways until September 22, 1953.

spaghetti plumped down on the earth just southwest of the intersection of Sunset Blvd. and Figueroa St.; in another spot, like the ruins of the Roman Colosseum—with the difference that the upthrusting columns are sharp-edged and plumb.

The destructive tornado trail was the future path of the freeway, and the spaghetti and Roman columns referred to the construction of the world's first four-level interchange. Those looking for human-interest stories about lives affected by property expropriation had to find them in other rags like the *Los Angeles Examiner*. The *Times* was criticized by some readers on this front. When the paper did address residents' plights, it was usually in the context of model citizens who didn't put up a fuss. Take Falvina Mascarina Bolluena, whose house was moved to make room for the Four Level. Despite her rain gutters getting knocked off and some other damage, Bolluena handled the disruption "with poise," according to the *Times*. Those who didn't were looking to "feather their nests" by holding out for more money, which "created pressing housing problems for many who had to move."

By 1950, most of the demolition and house-moving was complete. The State Highway Division estimated that 4,500 people were displaced between the Hollywood-Downtown corridor, a number that was regarded as low. (Rubble from old house foundations was trucked into Chavez Ravine, where it became fill material for Dodger Stadium.) Those houses that were moved were transported up to four miles away. E. F. Wagner, a right-of-way agent, observed that many of the homes "look surprisingly good after they move them. The owners usually remodel and redecorate when they move the houses," though he didn't specify who ponied up for those fixes.

A house being moved from the corner of Colton and Boylston Streets in 1948 to make way for the freeway. A woman stands defiantly on the enclosed front porch, presumably refusing to budge as workers jack up her house.

The state's bulldozers did manage to spare several landmarks. The route was rejiggered to avoid a new KTTV television station at Van Ness and Sunset (since torn down, now a high school). It also skirted the Hollywood Tower building and the First Presbyterian Church of Hollywood, accounting for the slight bend in the freeway near Gower Street. As new off-ramps were completed, cars were allowed access to these incremental extensions, though, per *Times* columnist Gene Sherman, "the younger set has been squeezing past the barriers and driving it for days." This would not be the last time that joyriding kids would flout unopened freeways.

THE FUTURE IS NOW

At 9:30 a.m. on April 15, 1954, Los Angeles prepared to snip another ceremonial ribbon of film. This time it marked the ten-mile completion of the Hollywood Freeway from Downtown

L.A. to the Valley. The project spanned eight years and came in at $55 million—$35 million over the original estimate—with expenses evenly split between construction and rights-of-way. The premiere-type event was led by perennial emcee Bob Hope. One thousand spectators lined the inbound lanes between the Mulholland and Pilgrimage Bridges as the comedian riffed on the near-decade wait.

"I'm going to miss my detour—Seattle is very pretty this time of year," the longtime Toluca Lake resident deadpanned about annoying street closures. "It took a long time to build this freeway, but the war interrupted construction. The Confederates captured it twice."

Up stepped the usual suspects to lay on more hokum, much of it hyping the Hollywood aspect of the eponymous freeway. The heads of the Hollywood Chamber of Commerce and Department of Public Works called it "the most famous freeway in the world" at a time when there were hardly any freeways in the world. Sheriff Eugene Biscailuz said it was "like an Arabian nights' dream come true," a nod to showbiz's dream factory. No, said Councilman Earle D. Baker, it was a "$55 million dream come true"—a figure that coincidentally mirrored MGM's movie

An iconic view of the freeway heading downtown, backdropped by the Spring Street Courthouse, Hall of Justice, and City Hall. This photo became a template for postcards advertising Los Angeles.

budget. Anticipating an uptick in auto accidents, Police Chief William H. Parker, who famously ran the LAPD like a military unit, urged drivers to use the freeway responsibly "so that the dream will not become a nightmare."

With the dreamy prelude complete, it was time for the main event. Several thousand motorists revved their engines behind barricades set up across all four southbound lanes, each eager to burn rubber over virgin concrete. "Here come the cars," quipped Hope as he cut the film strip. "Synchronize your bandages and jump."

The barriers were removed. The cars shifted into "drive." And then . . . they crept along, stuck behind a procession of plodding ceremonial vehicles. Instead of the Indy 500, it was more like the O. J. Simpson freeway "slow chase" forty years later, only slower. Witnesses recalled the chain of cars disappearing over Cahuenga Pass, backed up for miles deep into the Valley. No sooner did motorists finally start moving than they had to slam on the brakes near Vineland due to a stark-naked woman dashing through traffic. After a cop ran her down and wrapped a blanket around her, she was remanded to the psych ward at General Hospital.

"NUDE OUTSHINES FREEWAY OPENING," blared the next morning's headline in a Southland newspaper. As if on cue, the Hollywood Freeway—that perennial adversary of dull moments—had succeeded in stealing the spotlight from the bigger story.

While the freeway's curtain-lift enjoyed extensive coverage, gone were the dusty allusions to Gaspar de Portola and colonial pioneers of the overland route. Southern California, like the nation, had changed immeasurably since 1940. Leaps in technology

OFF-RAMP ↗

South of the Four Level, the Hollywood turns into the Santa Ana Freeway, which by 1954 reached Pioneer Boulevard in Norwalk. Consequently, the Hollywood Freeway's connection with Downtown enabled motorists to continue another fourteen miles— almost to the Orange County line.

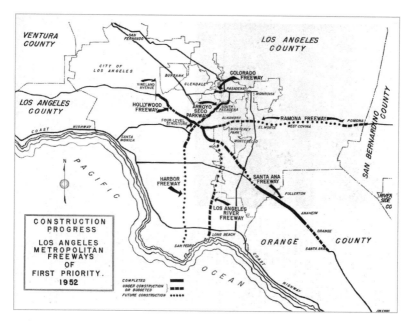

A 1952 map showing the progress of the Hollywood Freeway, including its northern terminus at Vineland Avenue. The broken-line portion between Vineland and the Four Level would be completed on April 15, 1954, finally linking DTLA and the Valley.

and a booming economy brought the Cold War to the cusp of the Space Age. America and the Soviet Union had secretly begun preparations to launch artificial satellites, and magazines like *Future Science Fiction* were all the rage. With the planned debut of Disneyland in the summer of '55, the portals of Tomorrowland would soon be flung open to all—just take the Hollywood Freeway to the partially finished Santa Ana and keep cruising south!

In a real sense, Tomorrowland had already arrived in Los Angeles. The Four Level was a huge hit, studied by engineers from around the globe and gracing picture-postcards. Its soaring overpasses and looping transition roads looked majestic from any angle—particularly overhead. At $5.5 million, "The Stack," as it was also called, was the most expensive half-mile cluster of roads in the world. It was also the first L.A. interchange to connect four different freeway names—the Hollywood, Santa Ana, Harbor, and Arroyo Seco.

This postcard signals the arrival of a hep, modern era—not just from the existence of the Four Level, but from the cool-cat writer inviting you to "dig" L.A.'s freeways.

As with Cahuenga Pass, excavators dug up storied history here. It contained the town's first gallows. Before that, it was Fort Moore Hill, a key site of the Mexican-American War where the first U.S. flag-raising occurred in California (a memorial stands on Hill Street, just off the freeway). Earthmoving tractors exposed bones and coffins from a bygone cemetery, prompting another wave of treasure-hunters to root around for gold. About all they found were old bottles and utensils. Despite its haunted past, there was no curse associated with the Four Level, though there was something equally onerous: a sulfur spring. The terrible stench permeated the construction site, which an engineer likened to working around rotten eggs.

It wasn't just The Stack that reflected new heights in engineering. Designers incorporated lessons from the Arroyo Seco Parkway into the construction of the Hollywood Freeway. "The Arroyo Seco was the First Generation," explained Caltrans veteran Heinz Heckeroth. "Hollywood was the Second Generation . . . really the first modern freeway." Shoulders were installed for emergencies. Lanes were widened. On-ramps were lengthened

Left: The Division of Highways piloted a variety of reflective markers on the state's highways. Right: Crewmen epoxying Botts' Dots to the surface of the Hollywood Freeway, March 6, 1966.

so white-knuckled motorists wouldn't have to rocket into speeding traffic from a dead stop.

Still, they didn't get everything right. Some access roads were placed too close together. A drainage flaw led to rainwater accumulating along the center divider; Bill Keene, the legendary traffic reporter, was among those who hydroplaned into the chain-link median. To discourage speeding, that old crank, Police Chief Parker, sentenced scofflaws who exceeded eighty-five miles per hour to five days in jail. He was even known to patrol the freeway himself at the speed limit, pulling over

OFF-RAMP ↗

If you've ever driven drowsily and then been stirred awake after the *thwump thwump thwump* of drifting into another lane, you can thank Elbert Botts. An engineer with California's highway department, Botts invented the white ceramic "turtles" glued to lane markings. Starting in 1954, Botts' Dots appeared as safety features on Southland freeways. But their days are numbered; Caltrans is gradually replacing the raised pavement markers with reflective painted lines as the dots get damaged or jarred loose.

anyone who passed him.

By the summer of 1954, the Division of Highways focused on the final stage of the Hollywood Freeway: extending it ten more miles, to the soon-to-be-completed Golden State Freeway in Sun Valley. The estimate came in at $20 million, a relative bargain that factored in lower rights-of-way projections, thanks to fewer landowners with whom to negotiate. Flush with young postwar families, the Valley represented the future more than the past,

With an exploding population, highway commissioners outlined plans for the Hollywood Freeway extension (today's SR-170) into the San Fernando Valley. P.O. Harding, a top engineer, promised relief from "the creeping paralysis of traffic congestion [from] existing street arterials."

progress over preservation. Still, there was pushback. When Valley Plaza Park was bisected by the freeway, a nearby homeowner accused L.A.'s parks department of "selling [the park] into slavery to the State Highway Commission, who allowed her to be permanently disfigured."

After the removal of several thousand homes led to more

By May 1972, the Hollywood Freeway truly had become, in Bob Hope's words, "the biggest parking lot in the world."

protests, the Valley Association for Freeways came to
the state's defense. Half of its
8,000 members worked for
movie studios. "We have no
more right to argue against
expert engineering findings
of the Highway Department
than did the 1,800 families
displaced when the [south-
ern] Hollywood Freeway was
built," wrote Chairman De-
Witt McCann to the simpatico
L.A. Times. "Our main purpose is to get freeways into the
Valley, which is sorely in need
of public transport routes."

OFF-RAMP ↗

By 1955, traffic had gotten so bad
that one incident spawned a new
medical term. Marion Grammer of
Van Nuys was so exasperated by
rush hour gridlock, the businessman stopped his car in the middle
of the Hollywood Freeway and sat
in a state of calm paralysis as cars
honked behind him. Police took the
catatonic motorist to a hospital,
where doctors diagnosed him with
"Freeway Neurosis" ("FN" for short),
a condition in which a person "whose
personal emotional problems and
the battle of the freeways add up to
a combination which seems too hard
to beat." Grammer was labeled "Patient One" (and, presumably, "Only").

The chairman had a point. In 1955, 183,000 vehicles a day
traveled between the Four Level and Vineland—double the capacity the thoroughfare was designed for. The city council explored
ways to alleviate traffic on "the biggest parking lot in the world,"
as Mr. Hope liked to say. One suggestion was to ban funeral pro-
cessions on the 101. After all, the thinking went, what was their
rush? Another proposal banned trucks. Neither measure passed.

One idea that did take hold was the placement of on-ramp
traffic lights, an experiment that showed promise in regulating vehicular flow. Initially, highway personnel actually hung out in the
bushes and controlled the signals by hand. This peekaboo system
eventually switched to automation, giving the Hollywood Freeway the distinction as the first freeway to employ ramp meters.

And yet. A persistent bottleneck continued to plague Cahuenga Pass, backing up rush hour traffic for miles. Engineers
blamed two archaic anomalies in the freeway's design that could
be traced back decades. The first pertained to Pacific Electric's

The white outline represents the proposed path of the Whitnall Freeway, which would have tunneled under the Hollywood Sign and popped out at the Forest Lawn cemetery in the Valley before swinging northwest. Much of the route exists today as Whitnall Highway.

trolley line. Abandoned after the Downtown extension, its right-of-way was now a weedy dirt patch teasing idle motorists. By the late '50s, the forty-two-foot-wide dead zone was converted to two automotive lanes—one going in each direction.

The other problem was a phantom freeway.

Ever wonder why the on- and off-ramps for Highland Avenue are three-lanes wide and include an extra-wide median near the Hollywood Bowl? At one point, planners envisioned the Whitnall (née Crenshaw) Freeway here. Its initial northerly route would have intersected the 101 along present-day Highland, and tunneled under the Hollywood Sign into the Valley. Highland's abnormally wide and long southbound exit was to serve as a connector.

Unfortunately, these extra lanes *stole* lanes from the Hollywood Freeway itself, which narrowed to three lanes going each way between Highland and Hollywood Boulevard. This accounted for the choke point. The Division of Highways eventually widened this segment. Although the Legislature eventually killed funding for the Whitnall, portions of it live on in North Hollywood, where grading had already begun. This resulted in the oddly angled, randomly broad boulevard to nowhere known today as

Still waiting for the Beverly Hills Freeway: The 101's extra-wide median where the north- and southbound lanes pass Vermont Avenue, facing south.

Whitnall Highway.

Whitnall wasn't the only freeway intended to rendezvous with the Hollywood Freeway. In building the Downtown spur, engineers separated the north- and southbound lanes near Vermont Avenue. The spacious median was intended to accommodate an exchange with the future Beverly Hills Freeway, whose path would have jogged north from Santa Monica Boulevard and connected with the Glendale Freeway (SR-2) in Echo Park. With a western beachhead in some of the ritziest real estate in Los Angeles, estimates for the route ran north of $300 million. Eventually, saner heads prevailed, and plans for the Beverly Hills Freeway were scrapped (though not until the 1970s). A Caltrans maintenance yard and storage units now haunt the specter of the interchange that never was.

Meanwhile, the Hollywood Freeway's tendrils kept creeping northward. On July 16, 1962, two North Hollywood booster

A ceremony marking the freeway's extension to Magnolia Boulevard, July 16, 1962. Old-time autos and oversized scissors had become almost tired props for ribbon-cuttings by this point.

committees sponsored the opening of the Magnolia Boulevard exit, extending the Valley terminus another 1.7 miles. The ceremony received yawning local coverage. Similar to the way Americans grew jaded with each subsequent Space Shuttle launch decades later, freeway fatigue seemed to be setting in. Or, more generously, so many freeways were being built that Angelenos simply started taking them for granted. Either way, the Hollywood Freeway's northern rollout highlighted the explosive growth of suburbia. The same day that the *Valley Times* covered the expansion, another article boasted that California's population would surpass New York's by Christmas. The Valley had a lot to do with this milestone. Its population had quintupled since 1945. By the early '60s, the San Fernando Valley—if it were its own city—would have been the tenth largest in the U.S., its 750,000 souls on par with the population of St. Louis.

Finally, almost six years to the day after the Magnolia opening—as the teenage pilot was crash-landing nearby—the City of Angels threw its final bash for the Hollywood Freeway. On July 3, 1968, the twenty-eight-year-old freeway got hitched to the Golden State Freeway. Already attached to the Santa Ana Freeway to the south, this completion was more than just symbolic. "It connects three L.A.s," observed Joel Kotkin, an urban studies expert. "Downtown, the center of commerce; Hollywood, the center of legendary L.A.; and the San Fernando Valley, the center of suburbia. It takes you through more worlds faster than any freeway in L.A."

The final tally for the freeway was $85 million covering 16.6 miles, with its northernmost miles traversing State Route 170. At the so-called Hollywood Split interchange, the northbound 101 becomes the Ventura Freeway and makes for the coast, while the 170 continues onto the 5.

SR-170 is a bit of an odd duck. Northbound traffic lacks connector ramps to the 5 South and the 134 East, while southbound lanes lack a transition to the 101 North. These annoyances—which momentarily force motorists onto surface streets to head in their desired direction—are the residue of a "Downtown-oriented design," according to the transportation blog *Let's Go L.A.* "This reflects the thinking of the era, that people would drive to a Downtown central business district in the morning and out to suburbs in the afternoon." However, ramps were always envisioned down the line. That's because the 170 was intended to be the northern stub of the Laurel Canyon Freeway, which would have disemboweled its namesake canyon and

connected with the 405 to the south. When dreams of a Laurel Canyon Freeway fell apart, so did motivation for the ramps.

FADING GLORY

Now more than a half-century old since its final dedication, the once-streamlined Hollywood Freeway has settled into turgid middle age. According to transportation data, the 101 South from the Ventura Freeway to the Harbor Freeway is the second-most congested roadway in the United States. Despite this, the only major engineering project the thoroughfare has seen in the last few decades has been the addition of a high-occupancy vehicle lane on the 170—the least congested portion.

Nonetheless, in keeping with its drama-queen reputation, the Hollywood Freeway has always found a way to stay in the picture. The fact that it blazes through Hollywood has something to do with this. As far back as 1958, an aspiring swimsuit model posed on a bridge above an overturned tanker truck. As photographers flocked around her, she was finally run off after being warned that the span might blow. A year later, drivers going through Cahuenga Pass spotted ancient Romans roaming the hillside along Universal Studios; they were filming a scene from Stanley Kubrick's upcoming *Spartacus*. In 1967, the same hillside doubled as a pasture for 1,500 grazing sheep, part of the shooting of *The Ballad of Josie* starring Doris Day.

Other incidents played like deleted scenes from *It's a Mad Mad Mad Mad World*: Motorists skidding and spinning through a river of BBs that had spilled out of a pickup. A massive traffic jam caused by drivers scrambling across lanes to pick up $7,000 worth of quarters that had fallen off a Brink's truck. A lion discovered in a stalled car. A CHP officer delivering a baby.

None left as long a legacy as the live chickens that, sometime in the early '70s, presumably escaped from an overturned truck near Vineland Avenue. Although most were captured, several got away and, following the edicts of nature, had more chickens. Generations of Valleyites have spotted their descendants darting through the brush. I've never seen them myself, but a fellow Toluca Lake Pony League dad who grew up near South Weddington Park told me he used to hear clucking in the wash behind his house and occasionally caught a glimpse of them. Rumors that the chickens crossed the freeway to get to the other side are unconfirmed.

Which came first, chickens or the freeway? Theories abound how they arrived on the scene, but the most credible is an overturned poultry truck from which several dozen escaped from smashed crates. By the late '70s, they were nicknamed "Minnie's chickens" after Minnie Blumfield, a retired senior who looked after them.

OFF-RAMP ↗

Two of the most well-known "ghost ramps" in the L.A. freeway system are visible on the Hollywood Freeway: The old Barham Boulevard Bridge off-ramp from the 101 North, and its on-ramp heading to the 101 South. The access roads were decommissioned in the 1980s, although the Art Deco-influenced 1940 bridge itself maintains historic status.

Like any diva, the Hollywood Freeway has undergone a slew of facelifts over the years. The city has spent considerable time and capital trying to beautify its driving experience via public art initiatives and other schemes. One of the most famous murals in Los Angeles history appeared on the side of an industrial building near the Edgeware Road overpass. Artist Kent Twitchell's *Old Woman of the Freeway* was a spellbinding portrait of a white-haired lady—modeled after actress Lillian Bronson—cloaked in an afghan billowing in the wind. The National Endowment for the Arts granted him $5,000 for the job. "There is such a contrast between this quiet old lady and the noisy freeway," Twitchell explained of his subject matter.

I was a callow eight-year-old when this mural suddenly materialized in 1974, and I am not exaggerating when I say its subject haunted my dreams. It was her eyes—they were a cloudy blue, like a blind person's, but one who could see into your soul. I also mis-

took the fluttering shawl around her neck to be a python (I thought the stitches were scales). Behind her, the night sky and bluish full moon (or is that Earth? Is she on the *moon*?) only added to my unease. In fairness to Twitchell, he did leave a lasting impression,

Artist Kent Twitchell's *Old Woman of the Freeway* mural, shortly after its 1974 debut. After it was painted over, Twitchell recreated it on the side of a building at Los Angeles Valley College.

Two of the seven hyperactive *LA Freeway Kids* depicted in Glenna Avila's mural along the Downtown Slot, one of several in this stretch painted for the 1984 Olympics.

which I suppose is the goal of every artist. In some ways, she personified the creeping senescence of the freeway itself.

Several years later, the city commissioned 2,500 murals in preparation for the 1984 Olympics. Fifty-one murals were painted along freeway walls, with some of the more memorable ones running through the 101's "Downtown Slot." Sadly, taggers blotched out most of them, despite a restoration effort in 2002. Even Twitchell's *Old Woman* didn't make it that long; she was painted over illegally by a billboard company in 1986, briefly resuscitated in 1995, and then buried for good in graffiti. To this day, driving northbound near the Edgeware overpass, I gaze up at her former wall canvas, and the adult I've become mourns her absence.

In the 1990s, the Hollywood Freeway played center stage to Los Angeles's inferiority complex. Back then, L.A. was considered a second-rate city among not only cultural purists, but also elected officials. Angelenos were told there were no landmarks to signify their hometown's importance to the world. Oh sure, there was the Hollywood Sign, but movies were fluff. Where was the city's Golden Gate Bridge? Its Gateway Arch? Its Statue of Liberty? Something that proclaimed L.A.'s melting-pot status as Ellis Island West?

Mayor Tom Bradley organized a committee to set up a "gateway monument" contest, open to submissions. It was determined that said artwork should be built over the entrenched Hollywood Freeway through Downtown. Out of 150 entries, the winning design was by a hot young architect named Hani Rashid (a New Yorker). All he would require to build it was $33 million in private donations—probably more, since the city would need to lease the air space over the freeway from Caltrans.

Rashid called his avant-garde creation *Steel Cloud*. Those who saw a mockup of the upcoming piece called it a broken Erector Set, and that was one of the nicer descriptions. It would essentially be a jumble of angled steel girders, and suspended among the beams would be a *Blade Runner*-esque mash-up that included a three-story aquarium, LCD movie screens, walking paths, gardens, plazas, and an electronic synthesizer converting the sounds of passing traffic into music—all towering up to 150 feet above the 101. And who better to push this futuristic vision than *Star Trek*'s Sulu—actor George Takei, who was a member of the committee? "The design says Los Angeles," Takei enthused. "It's energy. It's action. And it has to be over a freeway. Freeways are what we are."

"I'm not sure we want to monumentalize the freeway," countered Richard Keating, a Los Angeles architect.

"It's stressful just to look at the drawings," added Councilwoman Gloria Molina.

Caltrans also raised objections. Drivers would be distracted by the screens showing Hollywood movies. Not to mention the train-wreck element of it all.

The city received exactly one private donation: $100,000 from a Japanese firm that insisted the monument be changed to emphasize the immigrant theme that had been promised. The project mercifully died, but L.A.'s ambivalent feelings toward its freeways persisted. If freeways weren't meant to be monumentalized, they could still be beautified—maybe even sequestered, like a crazy old aunt retired to the basement.

Such a plan was presented in 2007, when the Hollywood

The Hollywood Freeway brushing past its namesake district and Capitol Records during the late '60s, before the addition of a center barrier.

Chamber of Commerce came up with an audacious proposal to bury the freeway from view. Hollywood Central Park would cover a one-mile corridor from Santa Monica to Hollywood Boulevards with a forty-four-acre green belt. Capping the depressed thoroughfare would allow the Hollywood neighborhood to weave back together after being torn asunder by construction in the 1940s and '50s. It would also serve one of the most park-poor districts in Los Angeles. Similar ideas were floated for the Downtown portion, as well as the submerged segment of the 134 Freeway through Glendale. By 2012, the project had the wind at its back. The Aileen Getty Foundation donated $1.2 million; coupled with the city's contribution of $825,000, an Environmental Impact Report seemed imminent.

Fast-forward to the present. The project has stalled and the price tag has soared north of $1 billion. Though supporters of Hollywood Central Park are actively keeping their vision alive, it looks as if the Hollywood Freeway will reside in a state of full exposure for the immediate future.

And maybe that's a good thing. While the 101's glory may have dimmed (the 405 is now the most iconic freeway in Los Angeles), there *is* something cinematic about cresting over Cahuenga Pass as one drives into the L.A. basin, especially during the so-called golden hour. On a clear spring day, the narrow pass flushes with viridescent chaparral. As the walls of the hillsides drop away, the incline toward Hollywood is a rush right out of a road movie, the distant skyscrapers of Downtown sprouting from the earth like giant monoliths. Then, there it is—the pulsating skyline to your right. The rooftop spire of the spherical Capitol Records Building. The flickering neon sign of the old Knickerbocker Hotel. Buildings old and new that resemble a studio backlot, backdropped by the finger-like Palos Verdes Peninsula and the gray shroud of the Pacific Ocean. The Hollywood Hills unfurl on your left in a perfect split-screen. Like an Italian coastal village, hundreds of homes in a mélange of architectural styles peek through the foliage (my eyes always latch onto Castle Ivar, which resembles Cinderella's Castle) capped by the Hollywood Sign and radio transmitter on Mount Lee.

As you graze past the Hollywood Tower—the model for Disney California Adventure's erstwhile Twilight Zone Tower of Terror ride—and the palm trees ringing the Chateau Elysée, you feel it. You are now enmeshed with the urban landscape. Onward through the Downtown Slot, you are bookended by the Plaza Historic District, the cradle from which this whole messy, marvelous city around you sprang, and the gleaming glass towers of old Bunker Hill. A harmonious collision of past and present.

Not a bad final act.

Hooray for the Hollywood Freeway.

THE HOLLYWOOD FREEWAY

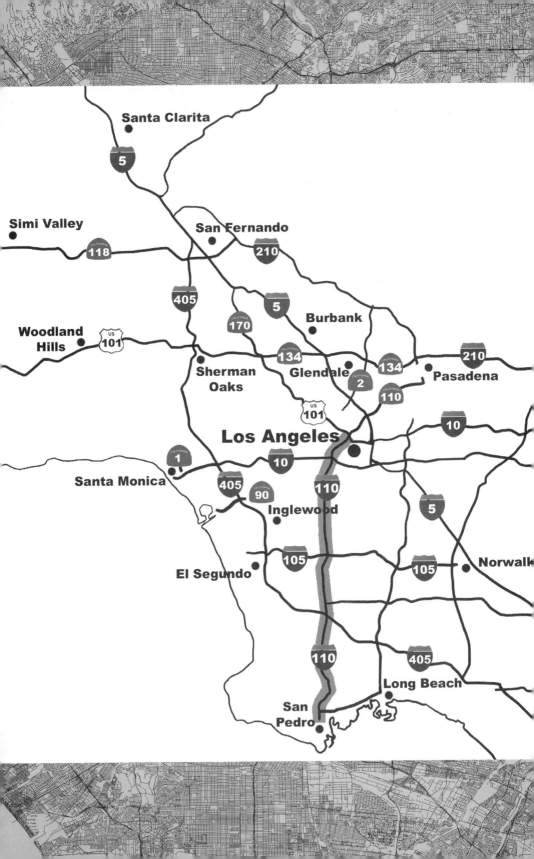

Chapter 3

THE HARBOR FREEWAY
Interstate 110
(1952–1970)

The Arroyo Seco and Harbor Freeways are both part of the "110 family," but these fraternal twins couldn't be more different. The Arroyo Seco (State Route 110) was designed as a parkway. Who can resist its curvaceous lanes, arched bridges, and Art Deco tunnels? The Harbor (Interstate 110), on the other hand, is the beast to the Arroyo's beauty. One must get past its ugly exterior to really appreciate it.

And that's okay. The Harbor isn't interested in preening for your pleasure. It's here to serve, the freeway equivalent of a thankless parent with teenage kids. It is the backbone between Downtown and the populous South Bay, and a key conveyance—along with the Long Beach Freeway—of cargo between Los Angeles and the shipping world. As a result, it ranks as one of the worst corridors for truck congestion. It also functions as a vital passage to LAX via the Century Freeway. This means that, at any given time, it likely carries an above-average number of cursing, wigged-out drivers worried about making a flight as they wage war with traffic. And forget about the poor saps who work in Downtown—you know, the ones whose teeth have been gnashed into powder from 110 PTSD.

In short, the Harbor Freeway is many things to many people. Perhaps this stress overload contributed to its reputation as the most dangerous freeway in America. Even solo drivers who pay for the privilege of rising above the fray in its so-called "Lexus Lanes" resent the Harbor Freeway. Sure, the smooth ride feels good now, but when the bill comes due, there's often a tinge of buyer's regret.

For such a thankless freeway, the Harbor had a strong champion in its corner: Vincent Thomas. You may have heard of him—something about a bridge. Thomas was the state assemblyman for the 68th District of San Pedro. Without his efforts, the Harbor Freeway, if it existed at all, might have been rerouted through Long Beach. That's where Kenneth Hahn comes in. The future county supervisor was a forward-thinker whose vision did not always align with Thomas's and his assemblymen. Hahn and Thomas are central figures in the story of the Harbor Freeway, their battle of wills a key plotline.

If you've ever examined a map of Los Angeles's boundaries—*really* examined it—you'll notice something peculiar. In silhouette form, it resembles a mosquito head. The southern border, jutting out of this head, looks like a long, narrow proboscis, its snout extending into the Pacific as if slurping up seawater. What accounts for this narrow corridor connecting Downtown Los Angeles with San Pedro?

The answer starts with a businessman named Phineas

Southerly view of the colossal Judge Harry Pregerson Interchange, where the Harbor Freeway crosses under the Century Freeway. In the distance, Long Beach Harbor and a portion of Catalina Island.

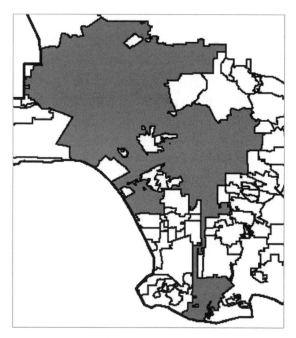

The Shoestring Strip—the narrow north-south strip of Los Angeles along the 110 corridor that serves as a gateway to the harbor.

Banning, who would come to be known as the "father of the Port of Los Angeles." Shortly after California attained statehood in 1850, Banning sank a fortune dredging a channel across San Pedro Harbor. The port had long been an anchorage to move goods—Spanish ships used to dock off the coast to deliver provisions to missions—but its shallow waters posed a hazard to the increasingly heavy freighters that visited SoCal's shoreline in the latter eighteenth century.

After completing his big dig, Banning went about building L.A.'s first railway: the Los Angeles and San Pedro Railroad. By 1869, it handled much of the trade between the port and L.A. In 1897, city officials declared the harbor the Port of Los Angeles.

There were just two niggling problems. For one, the Port of Los Angeles wasn't anywhere close to Los Angeles; the city's southern edge stopped near Downtown. Enter Gabor Hegyi, an enterprising developer who suggested that officials simply claim a strip of land between L.A. and the harbor as its own. So, on December 26, 1906, in a wildly generous interpretation of expropriation laws, the city acquired a sixteen-mile-long, half-mile-wide patch between Downtown and the harbor cities of Wilmington and San Pedro. The "dangling" parcel began to show up on city maps, where it was labeled the Shoestring Strip for obvious reasons. Years later, a portion of it would be called Harbor Gateway while the rest was absorbed by other communities.

But Los Angeles still had no way to access its port without going through the cities of San Pedro or Wilmington. Here the city had a fair amount of leverage. Its Board of Harbor Commissioners determined that, in order to safely convert to a full-functioning deep-water harbor, San Pedro Bay needed $30 million of improvements—an amount neither San Pedro nor Wilmington could spring for. In 1909, Los Angeles annexed both waterfront

A passenger steamship in the 1920s leaving San Pedro Harbor. Los Angeles hosted two luxurious steamers during that era, the SS Harvard and SS Yale, both of which made regular runs to San Francisco and could carry up to 800 passengers.

cities. Proboscis completed.

When the harbor over-
haul was finalized in 1924,
L.A. transportation officials
began widening Figueroa
Street to serve as the main
artery to the Port of Los Angeles, with a northern extension that reached Eagle Rock.
The work took nine years and
involved the excavation of the
four Figueroa Street Tunnels

OFF-RAMP ↗

Terminal Island was chosen as
the final assembly site for Howard
Hughes's *Spruce Goose* (so nicknamed because of the aircraft's
spruce and birch wood construction). The troop transport plane had
a record wingspan of 320 feet. The
billionaire pilot flew the H-4 Hercu-
les (its real name) only once, during
a test flight off Long Beach on November 2, 1947.

(later integrated into the Arroyo Seco Parkway). So began State
Route 6 (later, SR-11).

With the onset of World War II, San Pedro Harbor took on a
strategic role. The government established Terminal Island as a
shipbuilding and defense bastion. Adjacent Long Beach became
the West Coast base for Pacific Fleet battleships, housing a Naval Shipyard. As important bulwarks to the security of the United States, the government rushed a local freeway that would link
both harbor regions, paid for entirely by Uncle Sam.

LOS ANGELES VS. LONG BEACH

On April 26, 1945, harbor residents woke up to screaming headlines from the *Wilmington Press* that underscored the urgency
of the war effort: "PATTON ROLLS UNCHECKED." "YANKS
READY FOR BIG OKINAWA PUSH." "MILAN, GENOA LIBER-
ATED." On that same front page, in small typeset near the bottom,
were the words: "FIRST STEPS FOR HARBOR FREEWAY."
The embryonic Terminal Island Freeway, flush with $14 million,
would soon be ready to "prepare for the big Pacific push"—a huge,
flag-waving win for the citizens of San Pedro and Long Beach.

The 3.5-mile freeway was obsolete from the moment it

opened in 1947. The war had ended, and Long Beach was already planning its own freeway. This one would also start in Terminal Island and converge with outlying highways, rendering the feds' freeway useless. Like Hughes's *Spruce Goose*, the Terminal Island Freeway (SR-103) was a day late and a buck short. Consequently, it holds an ignominious record: lowest traffic volume of any freeway in the country, recently averaging a paltry 11,000 vehicles a day. Movements are underway to decommission the freeway's current 1.6 miles and turn them into a greenbelt.

The Terminal Island Freeway wasn't the only outmoded roadway serving the harbor. During the 1940s, tens of thousands of people either worked or lived in the South Bay, many drawn by oil, steel, aircraft, fishing, manufacturing, and shipbuilding jobs. The Bethlehem Shipbuilding Corporation employed over 90,000 people alone. Figueroa Street, it was now clear, was not wide enough to sustain future growth. Los Angeles City Council members added a Figueroa Freeway to their wish list for state funding, but they nearly got checkmated by their rival, Long Beach.

The Los Angeles River Freeway—later renamed the Long Beach Freeway (I-710)—would funnel traffic through Long Beach

Workers assembling F-86 Sabre planes at North American Aviation in the South Bay region. The plant manufactured more than 9,000 of the fighter jets, which saw heavy action in the Korean War.

Top: The Terminal Island Freeway facing the bridge over Cerritos Channel in 1948, with oil derricks outnumbering cars. This section is now simply the SR-47 Freeway.
Bottom: The freeway's intersection with Seaside Boulevard on Terminal Island. The boulevard was upgraded to the Seaside Freeway, which connects to the Long Beach Freeway (I-710).

and provide a more direct freight route to and from the rail yards and warehouses of East Los Angeles. To show they were serious, Long Beach officials pooled $1 million to begin construction. This was a bold maneuver that circumvented protocol; typically, municipalities would work with California highway commissioners on approvals, financing, design, rights-of-way, and construction. Long Beach saw it as a calculated risk, a kind of "If we build it, they will come" approach.

Indeed, as they rolled out their thoroughfare, city council members continually pestered the state for more money. A commissioner called out this unorthodox approach in a report to colleagues: "The southerly extension of the Los Angeles River Freeway now in progress by the City of Long Beach is the only instance since World War II of another governmental agency carrying out the construction and financing of a complete unit on the Los Angeles Metropolitan Freeway System."

OFF-RAMP ↗

In the late 1980s, Assemblyman Richard Katz proposed converting a portion of the Los Angeles River near Downtown into a freeway. His waterlogged idea was roundly rejected; the river subsequently rode a wave of conservation support that continues to this day.

In the end, the gambit worked. But more immediately, the Long Beach Freeway emerged as a pawn in the political gamesmanship played between Vincent Thomas and Kenneth Hahn.

In 1947, Hahn unseated the incumbent for the Los Angeles City Council 8th District. The twenty-six-year-old whippersnapper was the youngest person ever elected to the council. His district covered South Los Angeles, between Western Avenue and Alameda Street—smack dab in the middle of the

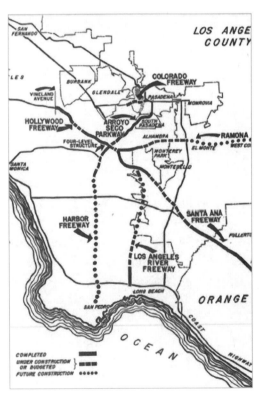

In the freeway-barren landscape that was 1953, the Harbor and Los Angeles River (later Long Beach) Freeways stand out as outliers. Both were locked in a competition to claim traffic between Downtown and the industrial ports.

Kenneth Hahn in front of a giant wall map depicting 1,500 miles of streets, highways and freeways in 1955. The supervisor points to the area south of Downtown that would be affected by the Harbor Freeway. County Road Commissioner Sam M. Kennedy squats next to him.

proposed Figueroa Freeway.

Hahn was a living oxymoron, a nerdy scrapper. He thrived on name-calling and once shoved a fellow council member back into his seat during a heated exchange. But he was also a deeply religious man who bled humanity. He pushed for more outdoor time at claustrophobic jails, fought for rent control, backed clean air, and even requested special protection for birds nesting at City Hall. Though he served only five years as councilman before being elected to the County Board of Supervisors, Hahn always looked out for his constituents in South L.A. as the specter of the Harbor Freeway, as it came to be called, hung over them.

As a councilman, Hahn felt that the future freeway should jog east at Olympic Boulevard, then travel along the Pacific Electric right-of-way near Alameda Street—similar to the path the MTA's A Line (née Blue Line) would take decades later. Such a route would minimize property acquisition costs and, more importantly, spare the destruction of his district. A study in March of 1948 found that simply to build the 1.5-mile segment from Temple Street to Olympic would require the removal of 1,824 residential buildings, affecting 15,921 people.

Hahn could only imagine what that portended for his much larger 8th District farther south. Not waiting around for any right-of-way agents, he took it upon himself to conduct his own study—from a hired blimp. He announced his blistering assessment in a series of press interviews and council meetings. "Asinine and ridiculous" was how he characterized the state's proposed freeway through South L.A. In fact, it was "the most stupid plan of the entire freeway program." He estimated that construction would require the removal of 30,000 homes of mostly low-wage earners. By eschewing the path along the PE right-of-way, the Division of Highways would be spending millions more to pay out homeowners. Hahn also asserted that the agency had reneged on a promise to study alternate routes.

Fellow councilman George Moore dismissed Hahn as a "nuisance" for going against the expert opinions of the state's best engineers. Despite winning public sympathy, Hahn was losing the battle. On April 3, 1950, Assemblyman Vincent Thomas of San Pedro introduced a resolution to green-light construction of the Harbor Freeway under its current plan. Three months

Protestors jam the gallery of a Los Angeles County Board of Supervisors meeting in 1947 after learning that the Harbor Freeway may gobble many of their homes.

later, the State Highway Commission held a public hearing on the adoption of the preferred route, and Hahn lined up dozens of homeowners to voice their opposition. Their grievances went beyond displacement. Many felt that the freeway was a waste of taxpayers' money better spent on national defense. As so often happened at these types of meetings, the homeowners were met with silent nods of sympathy.

Later that week, the L.A. City Council convened a vote on the freeway. Upwards of 200 citizens packed the chambers, turning the hearing into a series of heated contretemps that tested Council President Harold Henry's calls for order. Though Hahn's claim of 30,000 threatened homes had been downsized to 3,000, he still felt that his community was being "railroaded" and seemed to be seeking solidarity with his colleagues. The real enemy, he argued, wasn't himself. After all, he was never opposed to a freeway connecting L.A. and the harbor, just its routing. The problem was those ball hogs from Long Beach. Hahn now recommended that the route run along the L.A. River, which was, of course, the exact same course that Long Beach was charting for *its* freeway. Since the state allocated its highway budget on a year-by-year basis, perhaps Hahn felt that hijacking the river route was a clever way of beating Long Beach at their own game, in case there was only enough money for one of the projects to proceed.

At the end of three hours, the council voted on the original proposed route through Hahn's district. It was approved in a landslide: eleven to two. It would run twenty-one miles at an estimated $100 million. Alas, Hahn's last stand had ended much like Custer's.

FREEWAY FRENZY

With the all-clear from local and state officials, the Harbor Freeway trudged forward with all the efficiency of a Winnebago in the slow lane. Over the next two years, engineers paved only

Trenching the Harbor Freeway through Downtown in 1953. The northern section is visible in the upper left. Completed in 1952, it ran a mere six-tenths of a mile from the Four Level to 3rd Street.

six-tenths of a mile. But it was a sexy six-tenths. The Harbor Freeway's northern extremity was fixed at the graceful Four Level Interchange. To date, The Stack only carried traffic along the axis of the Hollywood and Santa Ana Freeways; the Harbor Freeway's perpendicular connection now made it 75 percent functional. The final node—the Arroyo Seco Parkway—would have to wait a few more months.

The July 30, 1952, opening of the Harbor Freeway's first stage was marked, of course, by a de rigueur ribbon-cutting. Attendees included County Supervisor John Anson Ford, highway commissioners, local transportation officials, and business groups. From the Four Level, the southerly freeway whisked over Temple, 1st, 2nd, and 3rd Streets, where traffic funneled off. When asked his thoughts about the four-block-long freeway, Hahn scoffed at its $16 million price tag. "Almost four million a block!" he said. Then, turning to his familiar scripture: "In the life hereafter, in Heaven, we are supposed to walk on streets of gold. But it seems that the bureaucrats are trying to build such expensive streets down here that we won't have to wait for the Heavenly Journey."

Bunker Hill, 1962: Angels Flight defiantly stays on track as residences and buildings are torn down around it. The historic funicular would close in seven years, re-emerging in 1996 half-a-block south on Hill Street.

The addition of the Harbor Freeway through western Downtown necessitated a reset of its street grid. Two-way roads became one-way. Some crossed the freeway, others converted to cul-de-sacs. A 735-foot tunnel—at $2.6 million—was dug under Bunker Hill to link 4th Street with the thoroughfare. Once a jewel of a neighborhood adorned with Victorian mansions, Bunker Hill had been left for dead by the '50s. The Harbor Freeway hastened its demise. Like other distressed localities bordering freeways, Bunker Hill was labeled as a "blight" that was impeding progress of the central city. Municipal planners dismissed the whole neighborhood as "half a century . . . of crime, disease, fire danger, and excessive public costs" that stood in the way of mostly commercial tenants promising a "1,000 percent increase in yearly tax revenue." By the end of the 1960s, 7,500

OFF-RAMP

In August of 1953, demolition crews knocking down a building in the path of the freeway unearthed a tube holding a copy of an 1893 San Francisco shipping guide. It was donated to the Marine Exchange in San Pedro, which has kept a record of ship arrivals and departures going back to 1923.

residents would be evicted. The *Los Angeles Times*, its headquarters mere blocks away, even casually suggested that the rickety old funicular known as Angels Flight would be replaced by an outdoor escalator. Only when all signs of the past were completely wiped out could Bunker Hill "regain its old majesty."

Meanwhile, south of Olympic, Hahn's resistance fighters threw sand in the gears of progress. The state was forced to file multiple suits to get title to eminent domain properties. In one case, the owner of a miniature golf course refused the highway division's offer of $283,000—a pretty penny in those days—by insisting that its fair market value was $600,000. After a twenty-three-day trial, a court awarded the pee-wee king $5,000 *less* than the initial offer. But it was a small consolation for the state. Expenditures on the Harbor Freeway were ballooning. Lengthy lawsuits, material shortages from the war, and a prolonged strike by AFL ironworkers all contributed to a $17-million overage in just the first year.

Frustrated with the glacial pace up north, Vincent Thomas turned up the heat in Pedro. A former professional boxer with deep roots in his hometown, Thomas fought hard for his constituents in the 68th District. But his style was more velvet glove than bare-knuckle. "He goes out of his way to make people happy," a family member said. San Pedroans loved him back, reelecting him to nineteen consecutive state Assembly terms covering thirty-eight years. In just the first few years of Thomas's tenure, San Pedro went from a backwater fishing village known for sardine canneries to a modern city sporting a new library, new fishing dock, and newly paved streets. A new freeway would bring a wave of fresh economic opportunities to its shores.

During Assembly sessions, Thomas saw that the uncomplicated Long Beach Freeway was getting more lovin' from the highway commission. And he wasn't about to wait for the state to grind through its quagmire of condemnation cases for buildings fifteen miles north of his district. He pressed legislators to speed up the right-of-way process in San Pedro in order to commence the southern terminus, stat. (He also tried to sell them on a bridge, but they weren't buying—yet. More on that later.)

With Thomas's persistence, by the summer of 1954 the state was able to clear a 2.5-mile swath between Battery Street and SR-1 (Pacific Coast Highway) for construction. As with other freeway projects, progress wasn't always measured linearly; rights-of-way determined which sections got paved first. The sight of engineers and earthmovers rolling in was a clear validation of Thomas's influence and a source of pride for the entire harbor region. For the groundbreaking ceremony on July 23, 1954, the assemblyman invited officials from San Pedro, Wilmington, Harbor City, and Carson.

Over the next four years, the Harbor Freeway began to coalesce. Press alerts went out with each new off-ramp as they came online: Olympic. Washington. Slauson. Almost all were accompanied by the kind of theatricality that defined the era. One of the dedications featured a shapely model named Ann Bradford,

Because lane changes are the cause of many accidents, the Division of Highways experimented with "Minimum Speed" requirements on the Harbor Freeway in 1966. The thinking was that motorists would find their most comfortable speed-category and "stay in their lane." The department's conclusion: "no positive advantages" but "definite disadvantages" as drivers largely ignored the signs. The program was scrapped.

who wore a sash emblazoned with the words "Miss Freeway Link"—certainly one of the clunkier female honorifics dreamed up by a Chamber of Commerce. Even the freeway's old nemesis, Kenneth Hahn, couldn't resist attending the 124th Street opening. At the ribbon-cutting on September 25, 1958, Hahn boasted that the freeway—now stretching ten miles—was already L.A.'s second-busiest after the Hollywood Freeway. When it's completed, he said, it will carry more traffic than "any street, highway, or freeway in the world."

The Harbor Freeway's immense popularity—even in unfinished form—did come with some growing pains for motorists. The Downtown section proved to be a confusing lattice of bridges and ramps that required quick lane changes and sudden start-stops. As anyone who has merged from the Hollywood Freeway onto the southbound Harbor Freeway can attest, the maneuver requires a "Frogger"-like thread of the needle through three lanes of traffic within a quarter of a mile, lest you find yourself involuntarily exiting one of the Downtown ramps. The nerve-racking exercise is compounded by incoming motorists from the Arroyo Seco crossing the same three lanes from the other direction—left to right—who are seeking the very exits you're trying to avoid.

Pulling off either move is nothing less than a navigational baptism for newbie drivers. Some drivers can't pull it off at all. Such was the case for Greg Morton, a thirty-four-year-old management consultant whose ordeal made him briefly famous. In March of 1958, just south of the Four Level, Morton attempted to weave to the right from the fast lane. Suddenly, a car veered into his lane and Morton panicked. He wedged the wheel leftward and found himself marooned on the center median, which, in those days, was simply a raised concrete strip with planters spaced every twenty feet. These planters posed a problem for Morton. He didn't feel he could get a "running start" to rejoin the stream of whizzing cars. So, he waited for a break in traffic. And waited. And waited. As he was stranded, he tried to flag down

A stranded motorist using a freeway call box in 1970.

eighteen passing police vehicles for help. Only one stopped. "You got yourself up there, didn't you?" the officer chided. "Just start your engine and drive off." Which is exactly what the cop did.

Things got so bad, Morton finally said to hell with it. He took a beach towel out of his trunk and started to sunbathe right there on the median. Perhaps this odd spectacle is what finally made a Good Samaritan assist this clearly delirious individual. The stranger was a civilian on a motorcycle who promised to make a call from a phone booth for help. Sure enough, a sympathetic officer arrived within minutes and stopped traffic long enough for Morton to escape the median. All told, the Highland Park res-ident had been stranded for an hour and fifteen minutes.

When asked about it later, Morton was shaken but took it all in stride. "I'd have given twenty bucks if, as there should be, there'd been a telephone out there I could've used to summon help," he said.

Perhaps Kenneth Hahn was listening. Four years later, Hahn—by then a county supervisor—was the driving force behind the installation of roadside call boxes. Hahn posed for a photo at

Aerial traffic reporter Francis Gary Powers piloting his aircraft above L.A.'s freeways for radio station KGIL in 1973. Tragically, Powers would lose his life four years later after crashing his helicopter while working for KNBC News.

one, placing an emergency call. It was on the Harbor Freeway.

While call boxes would have to wait a few more years, 1958 did see the first routine traffic reports from helicopters. Prior to this, freeway conditions were conducted by roving cars or sporadic airplane flights. Radio station KABC was first out of the gate with *Operation Airwatch*. Every weekday morning and afternoon, traffic jockey Donn Reed delivered rush hour updates from the cockpit of a Bell whirlybird. It was an instant hit with motorists, and Reed had proof. One morning, he asked any drivers who saw his copter to flash their headlights. Six out of ten cars did.

The fact that so many commuters tuned in may have saved the life of a three-year-old girl toddling through traffic on the Harbor Freeway. Reed got his studio to cut into programming so he could warn drivers about her presence. As cars slowed and paused, she wandered off the thoroughfare, no worse for the wear.

Not surprisingly, the Harbor Freeway saw the bulk of traffic updates. By 1958, more than 318,000 vehicles per day were passing through The Stack. That same year, the Dodgers kicked off their inaugural season in Los Angeles after relocating from

FREEWAYTOPIA

Dodger outfielder Ron Fairly poses with Long Beach mayor Ray Kealer in the cavernous climes of the Los Angeles Memorial Coliseum. The team would move up the 110 Freeway to Dodger Stadium for the 1962 season.

Brooklyn. Home games were played at the Los Angeles Memorial Coliseum as the team awaited their permanent field in Chavez Ravine. Built for the 1932 Olympics, the Coliseum's football-length field was not designed for baseball, just as its dense Exposition Park neighborhood was not suited for battalions of cars jamming its streets from spring until fall. Parking lots around the Coliseum could accommodate only 3,400 vehicles, forcing most motorists to pay to park on people's lawns or find street parking. One fan from Phoenix who flew in to catch the game had to walk twenty-four blocks afterward to find a taxi to his hotel—a longer journey than his plane ride.

Crushing traffic around the Coliseum backed up on the Harbor Freeway for a mile or more in each direction. The delays led to a Dodger fan stereotype that persists to this day: "Fans have been arriving as late as the third inning," pointed out sportswriter Rob Shafer of the *Pasadena Star-News*. Mostly, though, Angelenos were so enamored by their Boys in Blue, any inconveniences were met with wry wit. "The one thing the Dodgers forgot to bring with them when they moved to Los Angeles was the New York subway," quipped one newspaper. When Liberace had the gall to perform at the neighboring Los Angeles Memorial Sports Arena during a Dodger game, Rob Shafer swore that traffic on the Harbor Freeway created "some kind of human record for collective blood pressure."

When the Harbor Freeway wasn't making news with its epic traffic jams, it was drawing more unwelcome attention for its traffic accidents. We're not talking about the fender-bender variety. In 1959, an epidemic of gruesome incidents haunted front pages, reinforcing the adage "If it bleeds, it leads." Readers sipping their morning coffee were greeted by drunk drivers killing unsuspecting motorists, out-of-control big rigs maiming babies, a fleeing motorcyclist shot in the head by officers (his body soared seventy-five feet), and a mother who encountered "bad luck" when she tried to avoid a black cat crossing her lane, pinning her and her daughter underneath their overturned car.

And though it's rare today, back then it was common for out-of-control cars to fly off bridges. "HARBOR FREEWAY CLAIMS ANOTHER VICTIM: MAN DIES IN VIOLENT CRASH" blared one headline in the *San Pedro News-Pilot*. A driver doing eighty miles per hour on the Harbor Freeway breached the guardrail in

The Los Angeles Fire Department responding to a fiery freeway crash in 2019. For decades, the Harbor Freeway and its corridor communities have seen a disproportionate share of accidents and deaths compared to the rest of the city.

Wilmington. He was thrown from the car and fell fifty feet onto the street below. A stunningly graphic photo shows his bloodied, twisted corpse pretzeled on the pavement, his bloated face looking like a Michael Myers *Halloween* mask as two EMTs stand helplessly over him.

Sometimes fatalities related to the Harbor Freeway had nothing to do with car crashes. One man leapt to his death off the 6th Street overpass, landing in the fast lane. Another man slung a noose around his neck and hanged himself off the 54th Street viaduct during morning rush hour. His limp body, dangling over the freeway, led to several injurious accidents between horrified commuters. In one of the more bizarre cases, a female domestic worker stabbed a taxi driver as they cruised down the Harbor Freeway. The cabbie pulled over and expired, while his passenger was nabbed running down the freeway with a bloody knife in her hand. She had been worried he was going to steal $300 from her purse.

Exasperated by the spate of accidents, Superior Court Judge Howard D. Crandall mimicked Police Chief William H. Parker's zero-tolerance approach to freeway scofflaws. He sentenced two offenders to five days behind bars. Their crime? Driving eighty-four miles per hour on the Harbor Freeway. "The shocking toll of deaths and injuries means only one thing," he warned. "All drivers appearing before me on charges of traveling at excessive speeds on the freeways can expect to go to jail."

Perhaps because freeways still engendered awe, newspaper editors rarely explored the root causes behind the Harbor's horrible wrecks beyond speeding. If they had, they might have found the two-mile patch between the Four Level and Pico Boulevard to be awkwardly engineered. Maybe they could have calculated how many lives would be saved with the addition of center guardrails. Or called for a study on the merits of mandatory seatbelt laws to prevent bodies from catapulting through windshields.

Instead, freeway driving was simply seen as a recreational

In the late '50s, Kenneth Hahn ran point in fixing the freak show that was the Harbor Freeway. Citing its seventy-seven accumulative deaths, Hahn pushed for chain-link fences along the median. "I'm convinced that these center barriers will hold down the traffic death toll," he said. "Evidence of this is seen by numerous bulges in the chain-link fencing where vehicles slammed into the barriers only days or hours after they were installed." Fences along medians became commonplace, but safety-wise were not much of an improvement. Concrete barriers—initially known as K-rails—would eventually become the norm.

Chain-link fences were reinforced with lateral steel cables down the middle.

hazard. Problems flared up every once in a while like a bad head cold—or whenever a black cat crossed one's path. Even when driving safety measures were eventually implemented, the Harbor Freeway remained deadly. As late as August 1989, Caltrans's own data determined that it had a "50 percent higher accident rate over other L.A. freeways." The two miles south of the Four Level remain the most lethal stretch of freeway in L.A. County.

AN ELEVATED PROFILE

On the brink of the 1960s, 500,000 people called the harbor region home. The Port of Los Angeles was the busiest on the West Coast, and the Harbor Freeway was statistically the busiest freeway in the world (though it has since been eclipsed by others). As L.A. mayor Norris Poulson pushed the Division of Highways to pave the final five-mile gap in Wilmington as soon as possible, Vincent Thomas focused on his signature project: the construction of a bridge linking San Pedro with Terminal Island. Thomas's idea for the span—which would replace a plodding ferry—went back twenty years. Critics derided it as a "bridge to nowhere," but Thomas, foreseeing the growth of the region, saw it as the height of efficiency by creating a continuous link between the Harbor and Long Beach Freeways. Eventually, the state Legislature came around to his idea of "closing the loop" and approved the San Pedro-Terminal Island Bridge. But like other infrastructure projects, it had to wait in line for funding, still allocated on a yearly basis.

Meanwhile, the presumptive final piece of the Harbor Freeway clicked into place on September 26, 1962. The ribbon-cutting took place in Wilmington with a photo op that included a model dubbed "Miss Wilmington." She and Thomas held oversized scissors that spelled out "WELCOME WILMINGTON" on the blades. All told, the 22.6-mile Harbor Freeway came in at $103 million over ten years—an impressive clip, considering its rough start. The only thing that could've made it a perfect day for Thomas would be a simultaneous dedication for his bridge—

which was, quite literally, *his* bridge. The Assembly renamed it in honor of his tireless loyalty to the harbor region. But the bridge still wasn't ready, and minus this connection to the Harbor Freeway, the freeway itself was technically incomplete.

Not to be upstaged by Thomas and his bridge, Kenneth Hahn briefly stole the spotlight back. Responding to growing

Talk about *owning* it: California Assemblyman Vincent Thomas and his Vincent Thomas Bridge in 1978.

concerns about the lack of mass transit along expressways, he announced a bold plan to build a monorail over the Harbor Freeway. Hahn envisioned the orderly tracks all over L.A.'s freeways, but as the busiest route in the city, the Harbor seemed like a logical place to start. A monorail down its median could connect Downtown with not only the harbor, but Century Boulevard, where another train in the sky could connect with LAX and its 6 million annual passengers. (Years later, of course, the Century Freeway would include light rail down its median and famously skirt the airport.)

Talk of monorails had been kicked around since the turn of the century, when Joseph Fawkes promised his Burbank-to-L.A. line—an idea so outlandish, his contraption was mocked as "Fawkes's Folly." Hahn's version was borne out of a 1954 proposal by the Goodell Monorail company. For $40 million, Goodell promised capsules zipping along at ninety miles per hour, connecting Downtown and LAX. It would be clean, quiet, and of no cost to taxpayers. But monorails were a logistical nightmare that involved the cooperation of multiple agencies, including MTA and the state highway department, which managed the air space over freeways. Monorails also went against accepted aesthetics put into place by the city's founders. Typical public transit was dirty, noisy, and cluttered the sky. Fearing the city would become another Chicago or New York, policymakers regularly rejected them—as they did Hahn's folly.

Meanwhile, in the crowning achievement of a life devoted to civic service, the Vincent Thomas Bridge opened to traffic on November 15, 1963. After all the ceremonial fanfare—fireworks, Navy Phantom II jet flybys, a two-hour parade—the assemblyman himself paid the first toll: 25 cents (tolls were eliminated in 2000 after the bridge earned back its $21 million in costs). Over a mile long, the twin-towered teal span remains the third-longest suspension bridge in California, behind the Golden Gate and San Francisco-Oakland Bay Bridges. Frustratingly, it would be another seven years before a $5.8 million, 0.7-mile extension linked

Does he have a license for that thing? Governor Ronald Reagan mans the controls of a bulldozer for a 1968 groundbreaking that would eventually link the Harbor and Long Beach Freeways.

the Harbor Freeway to the Seaside Freeway (SR-47), which leads over the bridge to the Long Beach Freeway.

On July 9, 1970, the loop was completed, and at long last, so was the Harbor Freeway. In some ways, it now had two southern termini; southbound motorists could either exit Gaffey Street (the freeway's official end) or head east over the Vincent Thomas Bridge. The 1970 dedication barely made a ripple outside the harbor area, but the *San Pedro News-Pilot* made sure to hit the important points in their front-page headline: "OLD TIME AUTOS, PRETTY GIRLS HELP DEDICATE FREEWAY."

Twelve years later—almost to the day—another big event visited the Harbor Freeway, though it passed unnoticed by most people. On July 1, 1982, the freeway graduated from being a state highway to an interstate. Overnight, its route number changed from State Route 11 to Interstate 110 (though it took almost two years to swap out all 800 signs, which led to widespread confusion). More importantly, the new designation meant the interstate was now eligible for federal financial aid. Tapped with this flush source, the California Department of Transportation (Caltrans) targeted the Harbor Freeway as a candidate for its Freeway Transit program, designed to alleviate traffic on the state's most

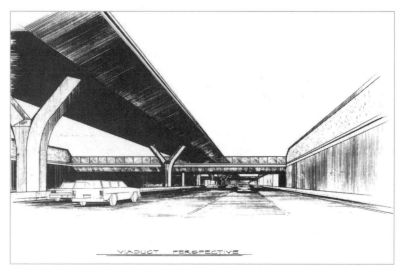

A Caltrans sketch of the elevated lanes above the Harbor Freeway. Explained engineer Ted Roworth: "The freeway was already 20 feet below grade and was hemmed in on both sides. You either go under the existing freeway or build a structure to carry the HOV lanes over the freeway."

congested roadways. Their solution was 19.6 miles of bus and carpool lanes rising fifty feet above the freeway median for three miles (the rest would be at ground level), which was projected to siphon 90,000 vehicles off the normal lanes.

With a push from Governor George Deukmejian, it appeared that the Transitway between the Santa Monica and Artesia Freeways would break ground by 1987. But the state overestimated the amount of federal highway funds it would receive for the $556 million project, and it was put on hold. The Loma Prieta Earthquake of 1989 added further delays; engineers had to prove to inspectors that the elevated lanes wouldn't suffer the same tragic fate as the Nimitz Freeway (I-880), which had collapsed and killed forty-two people in the Bay Area. Fortunately, the Transitway had already met a higher safety standard: it was built to accommodate light rail, should it ever be added at a later date. Finally, the Transitway opened on July 29, 1997. Though its carpool lanes were originally freely accessible, today they fall under the auspices of an electronic toll system—part of a broader movement toward HOT

(High Occupancy Toll) lanes and congestion pricing.

The introduction of L.A.'s first double-decked freeway was an impressive feat, but it was nearly overshadowed by its gargantuan junction with the recently built Century Freeway. If the original Four Level felt like a ride at Disneyland, the Harbor-Century Interchange was Disney World. It soared 130 feet in the air—the grandest, most expensive interchange ever built by Caltrans. One reason for its girth was the decision to give high-occupancy lanes their own connectors. But the glory of these soaring carpool flyovers lasted only a few short months before being tarnished by an act of unspeakable horror. As the *L.A. Times* succinctly stated, it was the ultimate L.A. story, with "guns, traffic jams, cellular phones, swarms of news helicopters, desperate self-promotion—and a sudden, tragic, cinematic conclusion."

At 3:00 p.m. on April 30, 1998, just at the start of rush hour, forty-year-old Daniel V. Jones parked his dark-gray pickup on the carpool ramp from the Harbor to the Century Freeway. In the cab with him was his dog, Gladdis. Passing motorists called authorities after a clearly agitated Jones pointed his shotgun at them. Police closed both freeways, resulting in miles-long traffic

The 110 Freeway at ground-level's Transitway Station, facing north. The split flyovers soaring in the distance are HOV lanes connecting to the 105 Freeway.

jams. As they spoke to Jones on his cell phone, he rambled about being unhappy with his health insurance. It was later revealed that he had cancer and had contracted HIV. Within minutes, TV news and law enforcement helicopters buzzed overhead. Jones exited his car and unfurled a large banner on the pavement that was easily visible from above, broadcasting this message into people's living rooms: "HMOs are in it for the money!! Live free, love safe or die."

Things quickly spiraled out of control when Jones began firing into the air. After he produced a knapsack, SWAT forces were called in as concerns were raised that he might try to blow up the bridge. Jones returned to his truck and ignited Molotov cocktails. A ball of fire blew out his back window and swallowed the cab, incinerating his dog. A flaming Jones burst out of his driver's door and rolled around on the concrete to douse his body, frantically stripping off his clothes. He wandered around in a half-naked daze before reaching for his shotgun. He anchored it against the guardrail, put its barrel under his chin, and fired—all caught live by seven television stations. As announcers uttered "Oh my Gods" and exasperated apologies, viewers watched Jones's body crumple to the ground across the banked carpool corridor.

Several networks had managed to quickly zoom out before his head exploded in a geyser of blood and brain matter. But others did not, and two of them—KTLA and KTTV (Fox L.A.)—had actually cut into children's afternoon programming to broadcast Jones's suicide. This incident served as a flashpoint that forever changed how local news handles high-speed chases or erratic behavior from the air. Camera operators now pull out to a wide shot if it looks like tragedy might ensue.

It would be cavalier to characterize Jones's suicide stunt as the culmination of the Harbor Freeway's role as L.A.'s deadliest freeway. Jones's plight could've happened anywhere. We'll never know why he chose this particular spot to publicly take his own life. One thing seems certain, though: he knew that his fiery spectacle and HMO banner, displayed in big bold letters to be read from the sky, would be catnip for ratings-hungry news stations. In this way, his public death was no different than the poor Harbor Freeway motorist whose pretzeled body had been plastered all over front pages forty years earlier after he was thrown off an overpass. We can't help but look.

If we take anything away from the Harbor Freeway, it's that it simply bears the burden of humanity and all its contradictions. Its majestic uber-interchange, venerated in the opening of the film *La La Land,* also staged a desperate man's incendiary send-off to the world. When it's not serving as a vital artery for freight trucks, it's contributing to growing inequity with its "Lexus Lanes." And the same freeway that obliterated Bunker Hill and parts of South L.A. unfurls like a welcome mat at the foot of a magnificent suspension bridge—a gateway to billions of dollars' worth of trade with the world. In the end, perhaps there is no better emblem of our sins and virtues than the Harbor Freeway.

Welcome to Los Angeles.

The Downtown Los Angeles skyline, as seen through the Harbor-Century Freeway Interchange.

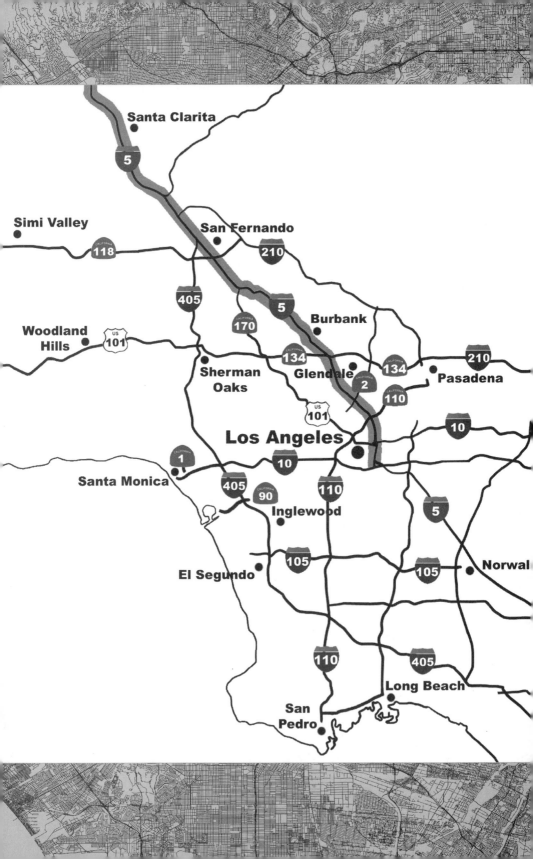

Chapter 4

THE GOLDEN STATE FREEWAY
Interstate 5
(1955–1974)

Is there a greater name than the Golden State Freeway? It not only conjures up ethereal riches and good vibes, but as part of Interstate 5—which runs the length of the state—it speaks to the diversity of the California landscape. The freeway itself was named after the old Golden State Highway (U.S. 99), a nod to the state's nickname. There's also something cool about a freeway whose descriptor is not a geographical place but a state of mind.

Despite these provocative connotations, the freeway is anything but. It doesn't have the classy curves of the Arroyo Seco or a seaside destination like the Santa Monica. The Golden State is a workhorse, that "get 'er done" friend with a pickup who helps you move in exchange for a couple of six-packs. Trucks, in fact, are one of its defining features. As a conduit between the Southland's ports and the rest of the state, the 5 Freeway is, depending on your perspective, a trucker's paradise or a commuter's nightmare. It sees more truck volume than any other Los Angeles freeway, accounting for almost one in four vehicles at the L.A. County/Kern County line. In the parlance of a Caltrans director, Interstate 5 is "the backbone of California's freeway system and our economy."

That's a lot to lay on one freeway, but the Golden State Freeway is tough enough. During construction, it had to overcome several high-profile showdowns known as the Battle of the Bulge and the Battle of the Barricades, not to mention a rifle-toting grandma. Once it was up and running, two separate earthquakes brought it to its knees. In one particular spot, it fell two different times.

As with many freeways, the Golden State exemplifies the dichotomy of Los Angeles. Created during an era of fallacious idealism—for white people, at least—it was an engine of commerce whose collateral damage included the

OFF-RAMP

Interstate 5 starts in Calexico, California, and ends in Blaine, Washington. Like its predecessor, US 99, it is the only interstate highway to reach the borders of both Canada and Mexico—a total of 1,381 miles.

homes, families, and parks of the Eastside. It's the California dream subverted, where the dream comes in conflict with reality, and reality is a bunch of people grinding through their lives, just trying to survive another day. Maybe that's why we all ultimately relate to the Golden State Freeway. Like that famous song goes, we get knocked down, we get back up again, but nothing's gonna keep us down.

FREEWAY

Bill Keene was the Chick Hearn of traffic reporters. During the same era that the legendary Los Angeles Lakers announcer was inventing phrases like "dribble drive," "slam dunk," and "no harm, no foul," Keene was KNX Radio's "eye in the sky," dishing out amusing advice to commuters to avoid bottlenecks. One area was so problematic, he had multiple epithets for it. It was "Malfunction Junction" or "The Beast." The *Redlands Daily Facts* referred to it as, simply, "Octopus." Other reporters called it "The Nickel/Dime"—the point where the 5 and 10 Freeways meet—or the "East Delay" Interchange, a play off its real name: the East L.A. Interchange.

I can think of a few other names I've personally called it, but one thing everyone seems to agree on is that the East Los Angeles Interchange is an impressively engineered monster. Its Dr. Frankenstein was Heinz Heckeroth, the same man who witnessed every stage of every freeway built in Los Angeles. Having cut

The southern hub of the East Los Angeles Interchange, where the Santa Monica, Santa Ana, Golden State, and Pomona Freeways rendezvous.

his teeth on L.A.'s early freeway system (the Arroyo Seco, Hollywood, and Harbor), Heckeroth was assigned the prodigious task partly for his deep knowledge of the city's traffic patterns and design standards. In those days, the Division of Highways built scale models of interchanges that it could show to construction companies—and just to show off. In a famous photo from 1958, Heckeroth poses next to the East L.A. Interchange as it appears in miniature form—an immaculate five-by-ten-foot model with serpentine freeways, street grids, and tiny trees, in which one inch equaled 100 feet. Standing behind his creation in a crisp suit and slicked-back dark hair, the dapper engineer looks like a young Mike Wallace.

Though only in his thirties, Heinz was undaunted by taking on the most expensive project in the history of his department, calling it a simple case of "just understanding the geometry" that resulted in thirty-two bridges. One aspect that makes it unique—and the busiest freeway exchange in the world, at 550,000 vehicles per day—is that, no matter which direction you come from, the motorist has the option of taking *three* different freeways. In fact, this mother of all interchanges gives birth to a quintuplet of freeways: the San Bernardino (I-10 East), the Santa Monica (I-10 West), the Pomona (SR-60 East), the Santa Ana (U.S. 101 North and I-5 South), and, just north of the interchange, the Golden State (5 North).

The Nickel.

Because it has to "share the road" with so many other freeways, the Golden State Freeway starts off as only two lanes through the interchange, creating a perpetual problem spot. (Rarely can one pass through the belly of The Beast at the maximum speed limit.) Coming out the other side, though, the 5 expands to four-plus lanes as it forges across the L.A. basin, largely mirroring old U.S. 99.

North of Santa Clarita, the 99 was

Top: Heinz Heckeroth lording over his creation, a scale model of the entire East L.A. Interchange, 1958. Below: Some surface-street signs in Burbank still display "Bakersfield" as the 5's control city, even though Caltrans switched it to "Sacramento" decades ago on overhead freeway signs.

Opened in October 1915, the Ridge Route was an engineering triumph that was nonetheless precarious, owing to dozens of narrow curves that hugged the ridgeline. Segments of the old road are still intact, thanks in part to Ridge Route preservation advocates.

Northwesterly view of the Golden State Freeway through Arleta, with incoming traffic merging from the Hollywood Freeway (SR-170).

known as the Ridge Route, a fabled but perilous highway with 1,500 curves ending in the San Joaquin Valley. Funds from the Collier-Burns Act of 1947 enabled this stretch to be widened from a narrow mountain pass into a four-lane expressway. Eventually, U.S. 99 was converted to freeway standards and decommissioned throughout the 1960s—absorbed by I-5—though the roadway does pop up again in Bakersfield.

The germ of the Golden State Freeway can be traced back to the 1930s. As one of the first freeways envisioned by planning commissions, it appeared on early maps under various names. The section between Downtown and the present-day Ventura Freeway was alternately labeled the Los Angeles River Parkway or Riverside Parkway. North of that, it was the San Fernando Parkway or San Joaquin Parkway. While the Hollywood

> **OFF-RAMP** ↗
>
> So what's the difference between a freeway and an expressway? Both are arteries for through traffic, but a freeway completely restricts access, with divided lanes and grade separations at intersections. An expressway has limited control of access; it may or may not be divided or have cross-traffic. In rural areas they sometimes intermingle on the same piece of road, which is marked by intermittent "Begin Freeway" and "End Freeway" signs.

Freeway would reach the Valley by 1940, the need for a second north-south freeway to handle the restless city was understood early on. Riverside Drive and San Fernando Road—thruways east of Downtown—had long since exceeded their capacities and were getting mangled by increasingly bigger and heavier trucks.

Several years before President Eisenhower's 1956 Interstate Act went into effect, the California Highway Commission was already anticipating a co-financing arrangement with Washington for this new freeway, since it was essentially an upgrade of a U.S. Highway. State officials broadcast plans to build the first two stages of the now-officially named Golden State Freeway. The first leg would run alongside Los Feliz, lopping off the eastern edge of Griffith Park. A southerly section would blow through Boyle Heights and connect to the future East L.A. Interchange. Commissioners coordinated with the Los Angeles City Council to sell their vision at public forums, no doubt expecting to be met with open arms amidst few questions.

But in what was now becoming a familiar scenario with each new project, the freeway wonks completely misread the room.

SAVE OUR PARKS

In multiple meetings throughout 1953 and 1954, hundreds of affected residents protested the adoption of the Golden State Freeway, and Griffith Park emerged as a major battleground. Given its origins, this was somewhat ironic.

During L.A.'s frontier days, the land representing the eastern flank of the Santa Monica Mountains was considered worthless. Sure, it offered perennial streams and some sweet picnic spots, but its steep, scrubby terrain discouraged home-building and lacked valuable natural resources. This did not dissuade Griffith J. Griffith, a walrus-mustached Welsh emigrant who had made a fortune in mining. In 1881, he bought Rancho Los Feliz, which included about 3,000 acres of the Santa Monica Mountains. Shortly thereafter, he took on a "Colonel" honorific, married into a wealthy family, and tried to ingratiate himself into L.A.'s high society.

But people found Colonel Griffith J. Griffith repulsive. "He's a midget egomaniac, a roly-poly pompous little fellow," remarked one associate. The Colonel would show *them*. In an 1896 version of a mic drop, he donated his entire acreage to the city as a Christmas present. Griffith Park was born.

Griffith J. Griffith in 1903, just before his incarceration for shooting his wife.

Unfortunately, the Colonel proved himself to be a raging alcoholic with a homicidal streak. During a 1903 hotel stay in Santa Monica, he drunkenly accused his wife, Tina, of conspiring with the pope to try to poison him. He then shot her in the face. She survived by crashing through an open window. Griffith did two years at San Quentin State Prison before returning to Los Angeles, where he attempted to repair his image with more overtures to public welfare. Not surprisingly, the city wanted nothing to do with him.

The Colonel died in 1919, leaving behind a trust that granted money to the city anyway—money that paid for the construction of the Greek Theatre and the Griffith Observatory in the years after his death. Both are now cherished institutions in a 4,310-acre park that attracts 10 million visitors a year. The repugnant Colonel finally achieved in death what he couldn't in life—love and adulation. Need further proof? Check out the giant statue in his honor at the park's Crystal Springs Drive entrance.

By the mid-1950s, Los Angeles Mayor Norris Poulson and Recreation and Parks joined the chorus of voices condemning the state's seizure of 200 acres of Griffith Park to make way for the Golden State Freeway. In the grand scheme of things, it didn't sound like much. But within that acreage, the parkland would lose its baseball diamond, playground clubhouse, pony track, miniature railroad, model airplane runway, archery range, six tennis courts, and nine holes from the fairways of the Wilson and Harding golf courses. In a letter to Paul Harding (no relation), the highway engineer overseeing L.A.'s freeways, Poulson and the Parks folks requested a reroute east of the Los Angeles River, an area of "dumps, cow pastures, and vacant land." Harding insisted that a westerly route had greater benefits, one of which was retaining San Fernando Road (old U.S. 99, on the east bank) as an industrial corridor. The state held firm—for now.

An even more dire situation was unfolding in Boyle Heights. Like Griffith Park, its Hollenbeck Park was a tourist destination. Early postcards show just how lush and pastoral it was, with a beautiful lagoon as its centerpiece. Boyle Heights was the city's most multicultural neighborhood, a harmonious melting pot that included Jewish, Japanese, and Latino families, many of whom ended up there due to the city's policy of redlining minorities. But the community began to fray in the '40s. Japanese families were ousted from their homes and sent to internment camps

Hollenbeck Park Lake, looking northeast, 1898. Talk about no respect: After the lower portion was reduced in size, the Golden State Freeway encroached on much of what was left.

The original right-of-way for the Golden State Freeway through densely populated Boyle Heights and points north. The Santa Ana Freeway (U.S. 101) branches off to the left toward Downtown. Hollenbeck Lake is in the middle.

during World War II. Jews moved to the Westside and the Valley as housing restrictions there began to ease. Appraisal agents for the Division of Highways characterized East L.A. as "infiltrated by minority groups, mostly [of] Latin derivation," helping drive down property values to the lowest in the Los Angeles area—under $3,000, versus Beverly Hills at $15,690.

Lacking a strong presence at City Hall, Boyle Heights was akin to a dying animal circled by vultures. The establishment of a statewide freeway grid—a principle soon codified in 1959's California Freeway and Expressway System—was given the highest priority. Routes were drawn straight through low-income neighborhoods, with little regard for the fraying of the social fabric. Virtually every decision was simple math, where people and businesses were assets inserted into a formula whose precept was: What serves the greater good?

Upon arriving at that decision, residents were relocated and businesses were bought out to make way for the freeway. In Boyle Heights that translated to 10,000 displaced individuals and the destruction of 2,000 homes, many of which were historic Victorians that Councilman Edward Roybal termed "some of the finest residential property" in L.A. The freeway would also drive a wedge through the heart of the business district. To add insult to injury, the lagoon in Hollenbeck Park was smack-dab in the freeway's path. Community activists responded with 15,000 signatures pleading for its preservation. While some of the park was saved, the present freeway literally soars over the lake's southwestern bank. Meanwhile, just one mile to the south, a flagship Sears department store—located in a gleaming Art Deco tower at 2650 E. Olympic Boulevard—had enough friends in high places to make sure the freeway would detour around it.

This landmark Sears building survived the 5 Freeway, but its luck ran out in 2021, when it closed after 94 years.

For the 5 Freeway viaduct over Hollenbeck Lake, the highway department promised "attractively designed . . . slender columns to blend in with the landscape." The result is a visually schizophrenic body of water: one end of the lake teems with ducks and fountains; the other exists in a state of perpetual darkness underneath a cacophony whirring engines and squealing airbrakes.

On the plus side, I-5's passage over Hollenbeck Park's lake makes it the only freeway in L.A. where one can theoretically fish off an overpass.

BATTLES, BULGES, AND BARRICADES

As the skirmishes in Los Feliz and Boyle Heights raged on, a third line of resistance was forming in the San Fernando Valley. Confronted with planning maps at town halls, homeowners from Pacoima, San Fernando, and Sun Valley tried to steer the freeway closer to the flood control channel that parallels Arleta Avenue (the state rejected their proposal). In Burbank, highway planners intended to divert part of the 5 along Glenoaks Boulevard, a major artery along the foothills of the Verdugo Mountains. Merchant and citizen groups responded with a "Save Glenoaks Boulevard" clarion cry. Taking their cue from a famous WWII campaign, they termed their war "The Battle of the Bulge"—a "bulge," in this case, that would be created by the proposed configuration.

The Burbank warriors took their fight to Governor Goodwin J. Knight, who punted on their cause. On March 29, 1954, the *Valley Times* ran an editorial imploring "all community forces" to wave the white flag, asserting that the proposed route "would do the most good for the most elements of Valley progress." Predictably, the state got its way here, too. Viewing an overhead map of the 5 today, one can easily spot the freeway's "bulge" between North Buena Vista Street and Sunland Boulevard, a 2.5-mile

An overhead map reveals the 5 Freeway's "Glenoaks Bulge," visible in the upper left as it deviates from San Fernando Road, the straight diagonal street north of Hollywood Burbank Airport's runways.

bump where it veers away from San Fernando Road and shadows Glenoaks before bending back to the former.

Meanwhile, as Boyle Heights was losing its fight against the freeway lobby, Los Feliz's hold on Griffith Park was tightening. An unlikely defender emerged in the ghost of Colonel Griffith J. Griffith. Well, okay, not his actual ghost, but his son, Van M. Griffith, who had inherited his father's air of haughtiness and uncanny knack for pissing people off. Van was appointed to the Police Commission in 1938, but regularly butted heads with Mayor Fletcher Bowron and his fellow commissioners. In 1946, the mayor had had enough, and several times asked Griffith to resign. Van wouldn't budge. "He either is unable to or refuses to work with other qualified members of this board," Bowron complained. The mayor finally just fired him.

Starting in 1955, Van Griffith began toting his heavy baggage to city council meetings. Now sixty-seven years old, his primary mission was keeping the park in his family's name untainted. The council welcomed his scrapper instincts. City Attorney Roger Arnebergh and Councilman John Holland teamed up with Griffith on a legal strategy that, they believed, could stop the state's

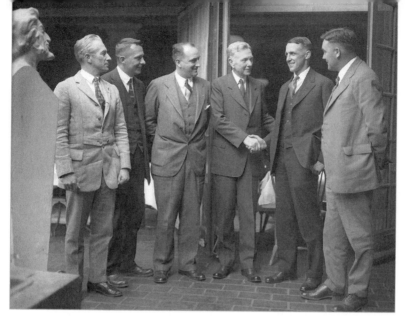

Park Commissioner Van Griffith, second from right, seen in 1920. A chip off the old block, Griffith J. Griffith's son was fiercely protective of his father's legacy in the fight against the freeway barons.

land grab of Griffith Park dead in its tracks.

In April of 1955, Griffith sought an injunction to stop the routing of the freeway through Griffith Park. His trial lawyer, Loren A. Butts, cited a reversion clause in Colonel Griffith J. Griffith's deed to the city. It expressly stated that the parkland would revert to the donor or his heirs—in this case, Van, the sole offspring—if it were used for anything other than recreational purposes. The deed even stated that any roads across Griffith Park must facilitate access to the park itself, not through it. Griffith's case was strong; a thoroughfare through the park appeared to violate the terms of the deed. Nonetheless, Judge Arnold Praeger denied Van's request to halt construction. Griffith vowed to appeal.

Praeger's decision seemed to embolden Frank B. Durkee, the highway commission's right-of-way kingpin. Well-versed in stall tactics by aggrieved parties, he proclaimed that the issue of re-routing the freeway was "done." But dissenters weren't done with him. Even as crews started to rip out trees and facilities, Larry Norman, president of L.A.'s Parks division, implored Durkee to idle his tractors and renegotiate contracts to divert the freeway east of the L.A. River. "The 4,500,000 people who yearly use the park are the sufferers," he said. The Griffith Park Men's Golf Club also refused to roll over. Their spokesman, David Jordan, collected 15,000 signatures to preserve the nine holes that would be lost to the superhighway.

As the fireworks continued in Los Feliz, the Division of Highways quietly made news nineteen miles away. On July 11, 1955, the first section of the Golden State Freeway opened in the largely rural northwestern San Fernando Valley. The three-mile stretch included a three-level interchange, which fused San Fernando Road (U.S. 99/U.S. 6) to the Golden State Freeway and several other arterials: Sierra Highway (U.S. 6), Sepulveda Boulevard (SR-7), and Foothill Boulevard (SR-118). The $3,301,500 junction was a major improvement over the dangerous street-level intersection it had replaced. (After the construction of the 210 and 14 Freeways, this interchange itself would be replaced.)

On Labor Day, CHP officers were thrilled to see traffic blasting through the exchange like water through a fire hose. It was a far cry from the previous twenty-five years, in which gunked-up traffic accompanied every holiday weekend. Not counting the forty-seven-mile Ridge Route—still in the process of converting to an expressway—the Golden State Freeway had 25.4 miles left to complete between this Three Level and the eventual East L.A. Interchange.

Shortly after the northern christening, the freeway's momentum continued farther south. On September 15, Van Griffith lost his appeal to reverse Judge Praeger's order allowing Interstate 5 to cut through Griffith Park. Judge A. Curtis Smith ruled that, under condemnation laws, the state's right "cannot be limited by contract of third parties, even including terms of the original bequest." In other words, state and federal interstate needs trumped local statutes. As already seen with the desecration of Hollenbeck Park, both judges' decisions were in keeping with 1950s transportation policies of prioritizing practicality, rapidity, and efficiency over quality-of-life concerns.

The Golden State Freeway in 1957, channeling between Riverside Drive (west) and the Los Angeles River (east) in Los Feliz. The trusses of the Hyperion Bridge, built in 1927, were just wide enough for the north- and southbound lanes to pass under a single arch.

David Jordan, the men's golf club mouthpiece, also suffered a setback, though the blow was more to his ego than anything else. After city council members hinted at making a deal with the state over the park's expropriated facilities, Jordan accused his one-time allies of shafting him by their "panic-stricken" overtures to Sacramento—an "indescribable blunder"—and threatened to initiate his own legal action to save the integrity of the golf courses. But what really seemed to irk him was a proposal by Councilman Harold A. Harby to close Griffith Park's greens to anyone residing outside the city limits. Jordan, who lived across the river in Glendale, was convinced the move was meant for him. (Given Jordan's self-righteous paranoia, could you blame the councilman?)

In the fall of 1955—with the worst of their obstacles out of the way—freeway crews finally got down to business constructing the $50 million, fourteen-mile leg between Boyle Heights and Burbank. One of their first tasks was the removal of what they termed an "unused small hill" in the northeast section of Griffith

Park; its 3.5 million cubic yards of dirt would make perfect fill material for grading.

But they would soon find out that everything comes with a price—even free dirt.

On January 6, 1956, a convoy of earth-removing trucks, having excavated all the dirt from the "unused hill," chugged over to a spot near Forest Lawn Memorial Park Hollywood Hills to continue mining another 200,000 cubic yards of earth. When they got there, they encountered a wooden barricade and two heavy-duty steel fences—six feet high, thirty feet wide—blocking access to the half-dug pit. Without warning, the barriers were put up at eight o'clock that morning under orders by Fred W. Roewekamp, the superintendent of park development.

"GRIFFITH PARK BARRICADE HALTS VALLEY FREEWAY" read the headline in the *Valley Times*, one of several newspapers picking up the story. Roewekamp took the draconian measure because the state's contractor—Vinnell Co. of Alhambra—was hauling away fill material without authorization. The contractor countered that they were simply "borrowing" the dirt. Moreover, Roewekamp's actions not only blocked crews from the pit, but also fenced in $300,000 worth of bulldozers,

Earthmoving vehicles back in business in Griffith Park after L.A. finally removed barriers that denied the state "free dirt." The Battle of the Barricades would stretch almost three weeks in January 1956.

earth conveyors, and other tractors left overnight. End result: the immediate stoppage of a 2.4-mile, $4.5 million section of the freeway northeast of Griffith Park. "The Battle of the Barricades" was on!

It seemed like a silly sandbox rhubarb on the surface. But frustration had been building within the Parks commission for months. Roewekamp's blockade was the exclamation point on a bigger beef—the cost of relocating Griffith Park's recreational facilities, or at least those that would survive the freeway construction (RIP archery range, model airplane runway, and baseball diamond). The state offered $400,000 in compensation. The city argued that that figure did not adequately account for reconstruction costs, and pegged the amount north of $2 million. The replacement of destroyed maintenance buildings would cost $1.5 million alone.

As the dirt embargo dragged on, the contractor grew anxious. Fifty employees were laid off, and $5,000 of daily income went up in smoke. And who was going to pay for the $1,200 penalty incurred for each day that construction exceeded the expected completion date? At one point, a Vinnell foreman considered knocking down the fence to access their imprisoned equipment, but decided against it for fear of upsetting the dead. "The only way out now is through the cemetery, and we can't do that," he lamented. Besides, even if the crews tried another entry, "they'd only put up another fence," the foreman said. One had to sympathize with the contractor, a figurative child caught up in a custody battle between two warring parents—the city and the state.

But the state held all the cards. Paul Harding, the engineer overseeing Southland freeways, blasted Roewekamp's churlish tactics. The highway department was more than happy to divert funds to Northern California freeways if L.A. wasn't going to play ball. This "inertia" could delay the Golden State Freeway for five years, he warned. This got the council's attention. Councilwoman Rosalind Wiener Wyman—who had recently engaged in talks with Walter O'Malley to move the Brooklyn Dodgers to Los Angeles—verbally reprimanded Recreation and Parks for its obstructionism. (I like to think she was motivated by Sacramento's threat to lavish attention on future San Francisco Giants territory.)

The barricades came down on January 23, 1956. Work resumed the next morning.

Still, bad blood carried on for months. The council drafted a resolution that banned contractors from "borrowing" any more dirt from Griffith Park without permission. Mayor Poulson accused highway officials of sandbagging construction as "punishment" for the city's opposition to the routing. Councilman Robert Wilkinson claimed $5 million had already been diverted to other freeways and bolted to Sacramento to make nice. Councilman Kenneth Hahn was so disgusted by the whole thing, he suggested blowing up the entire Parks division.

In the end, the city council and California settled on $1 million for the relocation and reconstruction of lost facilities, including two signature attractions that remain popular with families—the pony rides and the miniature railroad. Also restored were the back-nine of the Wilson and Harding golf courses. This led to one

Watch out for that 18th hole... it's got one heckuva hazard! These greens, hosting a tournament in 1937, have been paved over by the Golden State Freeway, as evidenced by Beacon Hill in the background. The promontory currently skirts I-5.

final turf war. In what must have been a slow news day, the *Valley News* reported that a clutch of unidentified golfers openly "glared" at Vinnell tractor operators puttering across the golf course, which "disrupted play for a short time."

You *know* David Jordan was one of them.

THE FINAL SHOWDOWN

On May 14, 1957, commuters who patronized Riverside Drive got a sneak peek of the future. With little notice, traffic was funneled onto a 2,000-foot-long section of the Golden State Freeway between Los Feliz and Glendale. By September 5, the entire 2.4-mile spur stretching up to Burbank finally opened after endless rounds of litigation and sabotage. In an extremely unusual move—but to the surprise of probably no one—there was no ribbon-cutting ceremony.

Construction continued to grind northward. By November, crews reached Alameda Avenue and built its $4.6 million cloverleaf access ramps. To the south, the finishing touches were put on Los Feliz Boulevard's partial cloverleaf, which ran $1.3 million. Right-of-way agents combed through Elysian Valley as earthmovers rumbled in. In a snapshot of just how prevalent freeways were on the L.A. landscape at this time, one unfortunate fellow was told by an agent that his house was in the pathway of the Golden State Freeway, and he needed to move. This greatly upset him. The man had only recently moved from another house to make way for the Hollywood Freeway.

But his ordeal was a mere footnote to a far bigger one that was transpiring in June of 1958. In the Silver Lake-adjacent neighborhood of Edendale, fifty-one-year-old widow Lomie Puckett was told to vacate her home at 2714 Gleneden Street. She refused. The offer on her house was too low. Claiming that "no one is going to push me around," Puckett parked herself on her front porch with a scoped thirty-caliber rifle, daring any freeway barons to take her home lest they wanted to eat lead.

Puckett quickly emerged as a folk hero. Headlines like "RIFLE-PACKING GRANDMA" and "SHOWDOWN ON THE FREEWAY" were splashed over mastheads across the country. In a series of photos as solemn as *American Gothic*, Puckett was a granite-faced portrait in granny glasses, a drab cotton house-dress draped over her large frame as she sat erect on her porch railing, cradling her firearm with an itchy trigger finger. The gun was no prop, either. It had belonged to her late police officer husband. "They ain't a-gonna pass," drawled the Texas native. "I can shoot the head off a rattlesnake at thirty paces."

ON GUARD — Mrs. Lomie Puckett, whose house is in the way of new Calif. Freeway, keeps guard with a rifle. While bulldozers worked only 50 ft. away. Mrs. Puckett sat on her porch, said she wasn't going to be pushed around. (AP Photo)

Lomie Puckett dominated national headlines for standing up to right-of-way agents. But like so many American heroes, there was more to this "granny's" story than meets the eye.

Lomie's adult son, Ross, also stood sentry, relieving his heat-packing mama when she needed sleep. She claimed that if the authorities took her away, Ross would take over. If they took him away, Ross's *friends* would take over. Media hounds didn't bother with Ross, since a twenty-three-year-old white man with a gun was a far less sympathetic figure.

Aiming to avoid a PR nightmare, the highway commission stood down, hoping the public would lose interest. But the story didn't go away. Puckett's defiance opened up a larger dialogue about the human price of progress. The *Tampa Tribune* uncovered other widowed old ladies whose lives had been heartlessly upended by superhighways. Grannies with guns seemed the inevitable end-point of a system in which people like Puckett too often "find themselves looking down the barrel of a surveyor's transit."

The standoff continued into early August, by which point 150 reporters were camped out front of Puckett's residence. Then, one day, sheriff's deputies posing as reporters accessed her property and handed her a piece of paper. It was a writ empowering them to seize her property. Lomie and Ross relinquished their rifles without incident. Within moments, bulldozers arrived on the scene, and photographers captured the moment they started devouring Lomie's house. But the white-haired Annie Oakley made it clear that this wasn't a surrender.

On October 21, 1958, Puckett took the state to court to get a fairer price for her house. The highway robbers had offered $8,250. She countered with $13,000. The parties ended up settling, but Puckett remained unsatisfied, complaining she was tricked off her front porch when Johnny Law served her with eviction papers. With her house now gone, she could no longer provide it as "evidence so the jury could examine it."

Proving that truth is stranger than fiction, the freeway dustup offered two twisty codas. It turned out that the Golden State Granny was far from destitute. She owned seventeen rental properties in the area, and wasn't even living in the one she claimed to occupy (it was between tenants). As for Lomie's co-armed defender of a son, Ross Puckett quickly moved on to a real job, and a pretty prominent one, too—he was hired by the Division of

OFF-RAMP

Two unbuilt freeways were planned as future junctions with the Golden State Freeway in L.A. One was the Chavez Ravine Freeway, a speedy bypass through Elysian Park that lives on as Stadium Way (and explains its wide berth). The other was the Colorado Freeway, its western anchor now the current cut-off for Colorado Street. Hence, the 0.6-mile stretch between I-5 and the first traffic signal on Colorado is built to freeway standards. An eastern spur lives on as the long access ramps for the Colorado Boulevard/Figueroa Street connection to the Ventura Freeway.

A vestige of the ill-fated Colorado Freeway's western terminus in 2015. Note the lack of a center barrier, which was added a year later.

Highways, Right-of-Way Department for the L.A. region. His job was to head up Evictions.

With federal dollars now flying in, construction on what would come to be called Interstate 5 ramped up at the end of 1957. Eighty million dollars was initially earmarked for the entire seventy-three-mile Golden State Freeway (a rising figure that was hard to track due to the piecemeal progress of the Ridge Route portion). Nine million was allocated to complete the southerly lanes to the eventual East Los Angeles Interchange, where the Golden State blends into the Santa Ana. Another $11 million was put toward the freeway's interchange with the then-called Pasadena Freeway. Pockets were filled between Burbank and San Fernando until, finally, the Golden State's northerly passage reached the San Diego Freeway.

On November 1, 1963, over 2,000 people flocked to the freeway interchange ribbon-cutting, which also marked the San Diego Freeway's completion in Los Angeles County. Four high school bands serenaded a gallery of beauty queens, honorary mayors, and old-timey car models, which, as with the Hollywood Freeway, represented the passing of eras up to the present.

Businesses were quick to capitalize on the hoopla. Tom Carrell's Chevrolet in San Fernando took out a full-page newspaper ad promoting a "FABULOUS FREEWAY PARTY." When they weren't checking out '64 Chevelles and Impalas fresh off the line, guests were invited to fete the Golden State's link from Downtown to the northern Valley. Carrell offered something for everyone: kiddie rides, fashion shows, free orchids "for the ladies," and, for the guys, Los Angeles Blades minor-league hockey players autographing pictures.

But business wasn't so great for many of Carrell's neighbors. Within months of the fusing of the freeways, motorists bypassed local surface streets. Merchants estimated that car traffic—a reliable gauge of the local economy—had dropped 70 percent, forcing some stores to close. One study in February 1964 showed that the number of daily vehicles on San Fernando Road was 5,181, down from 18,775 the previous year. Shifting traffic patterns caused by the opening of an interstate, of course, were nothing new; San Fernando was merely a microcosm for what was going on across America during the manic era of interstate-building.

One beneficiary of traffic reduction was the Hollywood Freeway. The addition of another north-south artery lowered its volume by 15 percent—much of it attributed to fewer trucks. Indeed, the generally flat Golden State Freeway became the preferred route for truckers, who no longer had to deal with the steep grade through Cahuenga Pass. Anticipating heavier truck usage, engineers of the 5 had wisely paved the two outside lanes thicker than the inner ones. By rerouting traffic away from the city's core, the 5 also relieved traffic through the 101's Downtown Slot, where daily usage dropped from 185,000 vehicles to 155,000 (of course, these numbers would swell again as the city kept growing).

Just north of San Fernando, however, the Golden State begins its first true incline. To keep traffic moving, engineers created separate truck lanes in 1965. They also continued to convert the expressway portion—from Newhall Pass to points north—into a proper freeway, a process that wouldn't be finalized until 1970. For local residents, improvements along Newhall Pass would literally be life-saving. The 5's unfinished state exposed several driving hazards when cars mixed with trucks—a steep grade, the lack of a center divider, road construction, street-level crossings, and occasional wet, slippery conditions in an area known to get snow flurries. One particularly treacherous spot was the intersection where the 5 met Sierra Highway (SR-14, formerly U.S. 6). In the first nine months of 1969, 47 people were killed in accidents, with another 853 injured—almost four deaths or injuries *a day*. Many casualties were the result of auto motorists slamming into plodding big rigs. But others were caused by trucks losing control and overturning or careening into oncoming traffic.

After years of carnage along the so-called Dead Man's Highway, the *Newhall Signal* unleashed a scathing editorial whose position was right there in its headline: "IF THE BIG RIGS WON'T SLOW DOWN, THROW THE BASTARDS INTO JAIL." Curiously exculpating somnolent middle-aged truckers, the newspaper blamed "overpaid young punks or overaged senior punks who dope along the freeway, lost in the fantasy that they are kings of the highway."

A solution seemed imminent. In 1969, the highway division announced its intent to

OFF-RAMP ↗

Two very weird accidents have occurred on the Golden State Freeway, one whimsical, the other tragic. Like a sendup of *Willy Wonka & the Chocolate Factory*, an overturned truck spilled $25,000 worth of liquid chocolate across its lanes in 1968. In 2015, twenty-year-old Richard Pananian raced up a berm on the southbound 5 through Glendale and lost control. As his car rolled over multiple times, his unseatbelted body flew through the air. It came to rest on the catwalk of the "Colorado St ¾ Mile" sign suspended over the freeway, horrifying passing motorists until firefighters removed the body.

The towering interchange ramps of the Golden State and Antelope Valley Freeways.

build a four-level interchange linking the Golden State Freeway with the Antelope Valley Freeway (the new SR-14). This would be no ordinary job for Kaiser Corporation, the contractor. At $24.6 million, it would rank as the most expensive project in the state highway system, a cost necessitated by the tricky engineering required to connect two freeways through a narrow canyon. The exchange would also include soaring connector roads and set another record to make acrophobes shudder: the highest concrete freeway span in the world at an estimated 150 feet—so high, one could imagine touching the sun.

Just like Icarus.

JUNCTION MALFUNCTIONS

There are obviously workmen outside our house, I thought.

My eyes had snapped open just before dawn to a low, loud rumbling. I immediately noticed that the walls of my bedroom were moving up and down. It was February 9, 1971, and I was almost five years old.

I wasn't panicked. I had matter-of-factly assumed that workmen were cranking up our house to work underneath it, maybe on some pipes. It wasn't until my mother came shrieking into my room that I realized something bigger had happened that I had never experienced before. "We just had a massive earthquake!" she gasped. "Are you all right?" I was fine, but seventeen miles away, it was quite another story.

The 6.6-magnitude earthquake had struck on a Tuesday at 6:01 a.m. It claimed sixty-four lives. Most of the deaths were from collapsed buildings in the Sylmar region. Just up the freeway, engineers were putting the finishing touches on the sky-high Golden State-Antelope Valley Freeway Interchange. It was 79 percent complete, and due to open within a year. That morning, the connector from the 14 to the 5 South came crashing down onto railroad tracks.

Meanwhile, a mile and a half south—at the Golden State's junction with the Foothill Freeway (I-210)—the crossover from the 210 West to the 5 South also gave way, crushing two men in a pickup truck. Other lanes buckled, fissured, or crumbled, including a truck lane for the 5. In all, a dozen freeway bridges were damaged from the quake, creating indefinite detours on side streets as forty miles of freeway shut down across the city.

Aerial shots of the gnarled interstates made them look small and frail, which made *us* look small and frail. Had we soared too high? Expected too much? Gotten too cocky? Within seconds, two of L.A.'s newest freeway labyrinths were reduced to rubbles of concrete and twisted steel girders, broken toys for some unseen giant's amusement. The crippled roadways resembled snapped blocks; the pancaked lanes, like they were stomped on by Godzilla.

When the dust settled, seismic experts determined that the doomed overpasses were far from earthquake-proof. The girders should have been reinforced. The columns should not have been set at dissimilar heights. The ramps needed hinge retainers. It would take $30 million to rebuild them right. More alarmingly, the broken Newhall Pass link exposed just how dependent the Santa Clarita and Antelope Valleys were on L.A.'s chain of freeways. "This is putting too many of our freeway eggs into one

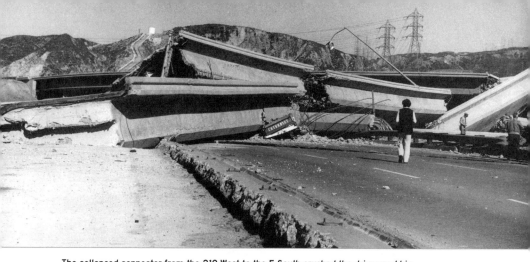

The collapsed connector from the 210 West to the 5 South crushed the driver and his passenger of this Chevrolet pickup truck, which was heading north during the Sylmar Quake. One of the occupants wore a wristwatch that froze at the exact moment of impact—6:01 am.

basket," observed Councilman Donald Lorenzen. "We have to have another way out of the Valley and that is all there is to it." He recalled that President Eisenhower's mandate for the federal highway system was to create safe means of access in times of national defense emergencies.

So was Lorenzen calling for a commuter rail line, à la the Metrolink system that would arrive two decades later? Far from it. His solution was, oddly, another freeway. The 24.3-mile Reseda Freeway was one of those projects collecting dust on a draft board. Similar to the Whitnall Freeway, it was a proposed route from the northwest Valley to the envisioned Pacific Coast Freeway via Topanga Canyon (Whitnall would've gone through Malibu Canyon, leading to its secondary name,

Overhead of the 210/5 Interchange after the earthquake. Note the gaping cavities in the Golden State Freeway. As a symbolic gesture, a sign now labels this section the Caltrans District 7 Fallen Workers Memorial Interchange, commemorating the approximately three dozen workers killed on duty in L.A.'s District 7 since 1921.

the Whitnall-Malibu Freeway). In fact, the Reseda Freeway was intended to be the southern continuation of SR-14. By Lorenzen's reckoning, a rebuild of the interchange seemed an opportune time to start this new spur to the ocean. But like those two other freeways, it never came to pass.

In April of 1974—a little more than three years after the quake—repair work on the Golden State-Antelope Valley Interchange was nearly done. It was time for its coming-out party. One dedication marked the ribbon-cutting of the Antelope Valley Freeway; another celebrated the completion of the Golden State Freeway from Boyle Heights to Kern County. One hundred and fifty spectators craned their necks in awe at the world's tallest interstate—intact and ready to thrill, like a roller coaster at the newly built Magic Mountain, just up the freeway. In retrospect, the only thing that marred the occasion was a jinxy comment by a low-level Assembly aide, who called the interchange "the safest yet."

On January 17, 1994—in circumstances eerily reminiscent of 1971's incident—another major earthquake struck the northwest Valley in the pre-dawn winter's hour. Centered in Northridge, this one had a comparable magnitude and fatality count, and caused even more widespread damage. Once again, the Golden State-Antelope Valley Interchange got bludgeoned. Two spans completely collapsed: the SR-14 South to the I-5 North, and the SR-14 South to the I-5 South—the exact same connector that had crumbled in 1971.

Because it happened at 4:31 a.m. on a holiday (Martin Luther King Day), traffic here was sparse and there were no

OFF-RAMP ↗

The partially collapsed Newhall Pass Interchange is memorialized on the cover of the Doobie Brothers' album *The Captain and Me*. The rockers, wearing Civil War garb, pose on a horse and buggy beneath a severed connector, while other photos show them picnicking on the half-finished freeway. When you call yourselves the Doobie Brothers, you can do whatever you want and it all makes sense.

This connector from the Antelope Valley Freeway to the Golden State Freeway North was one of two spans in Newhall Pass that gave way during the 1994 Northridge Quake. The Federal Highway Administration pointed to shear failure as a result of collapsing columns.

immediate casualties. However, about an hour after the temblor, LAPD motorcycle officer Clarence Wayne Dean was driving southbound on the SR-14 Freeway. Because it was still dark, he did not see that its ramp to the I-5 South was severed. He rode off the edge of the roadway and fell forty feet to his death.

The déjà vu aspect of the cursed interchange garnered a lot of attention, as did the downed bridges on the Santa Monica Freeway, the world's busiest thoroughfare at the time. Lost in the media coverage was the collapse of two eighty-foot-high viaducts just north of Newhall Pass, where the 5 goes through the Gavin Canyon area of Santa Clarita. Once again, a major north-south conduit between L.A. and the rest of the state was impassable, rendering about $550,000 in lost commerce per day. But there was some good news. Locals turned to the recently built Metrolink to commute into the city from the Santa Clarita and Antelope Valleys—an option not available in 1971. After the quake, North L.A. County ridership skyrocketed from 1,000 per day to 22,000.

Because it affected all lanes of the 5, repair work along Gavin Canyon received high priority. Governor Pete Wilson was among the backslappers on hand for its May 18 restoration. As a symbol of the freeway's economic importance to the state, tractor-trailers from Viking Freight System were invited to christen its newly opened lanes. The Newhall Pass connector was restored six months later, with a ceremony that included a jazz band from Magic Mountain. Reconstruction of both sections cost about $28 million, though the contractors pocketed bonuses for finishing the work early. In honor of the deceased police officer, the Newhall Pass exchange was named the Clarence Wayne Dean Memorial Interchange.

THROUGH THE GRAPEVINE

The Golden State Freeway ends its Los Angeles run in Newhall Pass. From there, it extends north for another thirty-seven miles. At the Kern County line, it simply becomes Interstate 5 and enters the Grapevine region, where it crests at Tejon Pass (elevation 4,144 feet) before descending into the farmlands of the Central Valley.

Grapevine travel advisories during winter storms hearken back to the Ridge Route days. Prone to icy conditions, its steep grades are often closed by the CHP, and it remains one of the most accident-prone roadways in California. When I was a kid, my aunt, uncle, and four cousins encountered black ice in the area on their way to visit my family from San Francisco. Their Volkswagen van rolled over two or three times. Not only did they live to tell the tale (many times at Thanksgiving dinners), but once the VW had been flipped back onto its wheels, my Uncle Irv insisted they keep driving. The van somehow managed to limp into our L.A. driveway with a crushed hardbody, shattered windows, and doors fastened with rope. In my mind's retelling, once they all exited the van, the whole thing spontaneously fell apart like the Dodge Monaco Bluesmobile at the end of *The Blues Brothers*.

OFF-RAMP ↗

The Grapevine is a community in Kern County named after wild grapes that grow in the area. The term has evolved to encompass the five-and-a-half mile grade in northern Tejon Pass—in some cases even referring to Interstate 5's remaining descent into the San Joaquin Valley.

My own experiences with the Grapevine aren't as dramatic, but one was strangely freeing. As I drove north on I-5 during the brutal winter of 1983, Caltrans shut the freeway down due to snowy conditions. Thousands of cars were suddenly stranded until it was safe to pass. As we waited, a Nerf football game broke out across the freeway lanes among us motorists, a woman handed out jerky from her pickup, and a big rig blasted Def Leppard. Total strangers became fast friends. When the roadway cleared, I never saw them again. But for a memorable hour or so, there at the end of the Golden State, we had all achieved one of our own.

Top: Yours truly as a teenage driver (with my cousin Sandy) experiencing my first freeway lockdown due to snowy conditions through the Grapevine. In addition to commiserating with other motorists, I witnessed several accidents that day, including overturned trucks. Bottom: A modern-day message board advising of snow where the Golden State Freeway ends—a mere hour's drive from where it begins at the frenzied East L.A. Interchange.

Chapter 5

THE FOOTHILL FREEWAY
Interstate 210 / State Route 210
(1955–2007)

Pity the poor Foothill Freeway. Its first segment rolled out within in a month of the Golden State's. Both were slated to open in their entirety in the early 1970s—two of SoCal's longest freeways.

But just as it was closing in on completion in L.A. County, the Foothill Freeway was plagued with problems almost biblical in scope—floods, mudslides, an earthquake, even smog. Its biggest obstacle was man-made: conservationists who blocked the state from cementing a "missing link" until their concerns were met. This led to years of delays that partly led to the retirement of an exasperated councilman.

Really, the Foothill Freeway was a victim of circumstances. Like Rip Van Winkle, it awoke one day to find that the world had changed. During its nascent years, the freeway lobby's guiding philosophy was "No neighborhood shall be more than four miles from a freeway." But environmental and social movements in the late '60s and '70s led to fundamental shifts in the ways we viewed freeways. Once celebrated, the in-progress Foothill was now maligned, an unnecessary and unwelcome intrusion through the peaceful Crescenta and Tujunga Valleys. Its Pasadena spur was also a lightning rod, resented for its role as a feeder to I-710—perhaps the most contested freeway segment ever proposed.

Nine years behind schedule and $24 million over budget, the Los Angeles portion of I-210 limped to the finish line in 1981. While only 18 percent of its eighty-six total miles lie within L.A. city limits—the primary focus of this chapter—it is an essential 18 percent that chains the Valley to Pasadena and the Inland Empire.

The Foothill Freeway also figures prominently into Los Angeles's narrative. Almost ten years to the day after its L.A. consummation, this transit afterthought was thrust into the limelight when it served as the set-up to the Rodney King beating at the hands of LAPD officers. To this day, the words "Foothill Freeway" and "Lake View Terrace" are linked to one of the darkest chapters in American race relations. Combined with its role as an ecological linchpin, the Foothill Freeway's 15.5-mile portion through L.A. has had a surprisingly outsize influence on the city's history.

A history that also lives on with Frank "Ponch" Poncherello's million-dollar smile . . .

Downstream from the San Gabriel Mountains, in a mesa just south of Jet Propulsion Laboratory, is the Devil's Gate Dam. Given the Foothill Freeway's bedeviled existence, it makes sense that it began its life here. Not many people are familiar with this

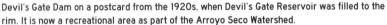

Devil's Gate Dam on a postcard from the 1920s, when Devil's Gate Reservoir was filled to the rim. It is now a recreational area as part of the Arroyo Seco Watershed.

The western flank of Devil's Gate Dam, back when it supported car traffic. This site was "ground zero" for the Foothill Freeway, which would obliterate the country road to the left to provide safe passage through the Arroyo Seco's narrowest section.

A sketch of the buttressed freeway spanning the gorge.

dam (its face is partially obscured by Oak Grove Drive), but when it was built in 1920, it was considered an engineering wonder that graced postcards from Pasadena. The name Devil's Gate refers to a craggy chasm below the dam that, some say, resembles the face of Satan. There is a long association with the dark side here, ranging from Native American and occult rituals to a culvert leading to the Gates of Hell.

Hell comes in many forms, of course. The 210 originated near the first dam built by the county's Flood Control District. Ironically, the freeway would almost be derailed by the same agency fifty years later.

A notorious danger zone spawned the freeway. Driving through the glen between Pasadena and then-Flintridge required deft navigation of narrow winding roads. "If you haven't been caught in it," the *Los Angeles Times* groused, "just try to imagine what it's like on New Year's Day a half-hour before it's time for the Rose Bowl tussle to begin." The remedy was a $2.2 million, four-lane freeway that would safely span the steep Arroyo Seco.

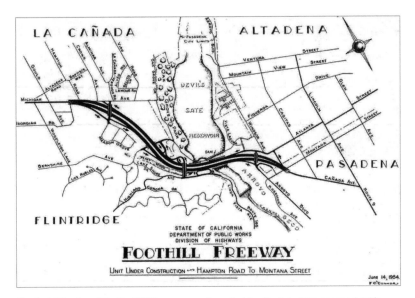

The first 1.8 miles of the Foothill Freeway, southwest of Devil's Gate Dam, mapped out in August of 1954. Portions of it would open by the following August.

Fully opened by October 1955, it was met with widespread approval even though it was only 1.8 miles long (and would later be reconfigured). It was an auspicious start to a long-gestating vision.

The Foothill Freeway had been on the drawing board since the 1940s. The general idea was to mirror Foothill Boulevard, a sixty-mile roadway that traverses the foothills of the San Gabriel and San Bernardino Mountains and includes much of historic Route 66. In 1958, the freeway was commissioned as an interstate with the recent passage of the Federal Aid Highway Act. At the time, the 210's eastern point was fixed at Interstate 10, where that freeway passes through Pomona. This terminus would shift northeastward over time. Like the Golden State (I-5), the Foothill wouldn't get assigned its interstate number until the Great Renumbering of 1964. Until then, sections of it still went by their original highway designations, such as SR-118 or U.S. 66.

After its nub near Devil's Gate opened, impetus for the 210 swung to the San Fernando Valley, where preparations for its link

OFF-RAMP ↗

So why is Interstate 210 called by those numerals? And why is it three digits instead of one or two? First, east-west routes end in even numbers. (North-south routes end in odd numbers.) Secondly, three-digit interstates are offsprings of their two-digit "parents." Put another way, Interstate 210 is a bypass off the much longer Interstate 10, which traverses the entire country. Also, because the first numeral is even, that means the 210 connects back to another interstate (I-5 to the west). Odd-first numbers, like I-710, do not. Decoding digits can be fun in a Morse code kind of way . . . or maybe it's just me.

with the Golden State Freeway were underway. In January of 1957, property owners from Sylmar, Sunland, and other foothill communities attended a three-hour meeting with state officials to learn about the project. Sipping coffee and chomping pastries, homeowners clustered around an aerial map to inspect a pending $14 million patch from I-5 to Hansen Dam. In addition to providing safer travels, the drive time for this 9.6-mile stretch would be cut in half, from twenty minutes down to ten.

Despite mostly positive grades for the freeway, there were murmurs of concern about how it would affect Big Tujunga Wash, an ecologically sensitive channel out of the San Gabriel Mountains that was prone to flash floods. Highway engineers George Langsner and A. L. Hutchison pressed the need to act quickly—niggling issues like flood control could be figured out in

The state's proposed path for the Foothill Freeway through Hansen Dam, starting with SR-7 (today's I-710 stub). Before it was recognized as I-210, the Foothill Freeway co-opted the SR-118 designation from the stretch of Foothill Boulevard that ran from the Valley to Pasadena.

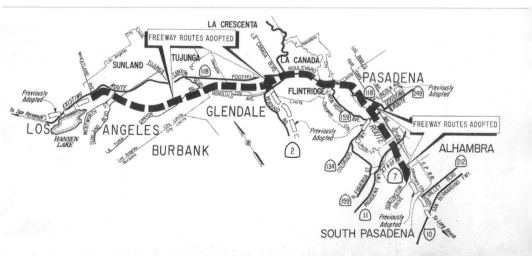

Seasonal mudslides and flooding were all-too-common occurrences in the foothills of the San Gabriel Mountains. This man is valiantly attempting to unearth his car in La Crescenta-Montrose in 1934.

time—employing scare tactics often used on interstate projects. Delays would result in further development of Valley suburbs, which would lead to higher construction costs due to increased land values, inflation, and more layers of red tape. More importantly, because the Foothill Freeway would rely on 90 percent federal funding, the reps explained, the state was under pressure to "use it or lose it." The Valleyites were in accordance: "Use it."

Meanwhile, progress on the 210 continued apace east of Sierra Madre, but hit the skids in more affluent neighborhoods like Pasadena and La Cañada. Set to conjoin with the Golden State Freeway in 1972, the Foothill Freeway started falling behind. At a hearing in Glendale's Civic Auditorium in 1964, the 3,000-seat room was divided into two opposing schools of thought. One side, led by Warren Dorn, chairman of the Board of Supervisors, agreed

Big Tujunga Reservoir and Dam in the San Gabriel Mountains. Completed in 1931, the flood control project is a vital bulwark for communities living in the floodplain of Big Tujunga Canyon.

with Assemblyman Frank Lanterman's assessment that Foothill Boulevard could handle traffic for several more decades. Both men offered a simple solution: rather than having the 210 West jog northwest from the Rose Bowl, why not just let traffic continue along the proposed 134 West to the 2 North? Sure, it was a less direct path to Sylmar, but only slightly

Two sons of Pasadena—Warren Dorn and baseball icon Jackie Robinson—in 1964. As a county Supervisor, anti-smog crusader, and self-proclaimed asthmatic, Dorn vigorously opposed the Foothill Freeway. When it was built anyway, it passed within three blocks of Robinson's childhood home.

so. La Cañada's school district also fell into this camp. It projected 6 percent less revenue from the removal of taxable properties. It also wanted to preserve La Cañada Elementary School, which would have had to be torn down and rebuilt elsewhere.

The other side of the room was represented by Glendale mayor Herman E. Barnes, whose jurisdiction included the Crescenta Valley. Barnes accused Dorn of self-interest in opposing the freeway's adoption: "We are more than a little amazed to find that Warren Dorn has seen fit to be a spokesman for his La Cañada neighbors and advocating the moving of the Foothill Freeway down to Route 134." He beseeched the highway spokespersons not to fall for "local hysteria."

Nothing was decided at the meeting. The *L.A. Times* agonized, "MULTI-MILLION DOLLAR QUESTION: WHERE WILL THE FOOTHILL FREEWAY GO?" It was a reference to the $320 million in government aid the state would forfeit if I-210 was not completed by 1972. In the end, the prospect

OFF-RAMP ↗

Sandwiched between the San Gabriel Mountains and San Rafael Hills, La Cañada and Flintridge remained separate cities until 1976, when the neighbors partnered up to become La Cañada Flintridge.

Driving along the 210 West, motorists enter a short tunnel under the La Cañada Boulevard overpass. Topping the tunnel is Memorial Park, replete with a gazebo, play facilities, and expansive lawn. It is considered a shining model for other proposed "cap parks," or lids, over trenched SoCal freeways.

of stalling was just too ominous to ignore. By 1966, the "northwest passage" from Pasadena to Hansen Dam was commissioned. La Cañada Elementary School would indeed come down, but in a compromise, engineers trenched the 3.6-mile stretch through that city and enclosed the section under Foothill and La Cañada Boulevards as a tunnel, upon which sits a park.

OFF-RAMP ↗

Just south of the Rose Bowl, the Foothill Freeway feeds the Long Beach Freeway (I-710, originally SR-7). This three-quarter-mile spur ends in Pasadena, though it was intended to pass through historic South Pasadena to connect with the main trunk of the 710. Caltrans acquired 460 properties along the proposed path, with another 2,432 living units threatened at one time. As recently as 2017, L.A.'s Metro board entertained closing the 4.5-mile freeway gap with a $5.6 million tunnel.

NATURE CALLS

On May 24, 1968, news photographers captured a bald man with black-rimmed glasses—the spitting image of TV's Phil Silvers—shattering a champagne bottle over the blade of a bulldozer. His name was Louis R. Nowell, and as the councilman for the northwest Valley, he was invited to commemorate the groundbreaking for the 210's western front. That same week, crews broke ground at its eastern terminus near Pomona, 48.5 miles away.

These celebratory symbols of civic progress belied the signs of the times. This was 1968, after all, one of the most tumultuous years of the twentieth century. One month earlier, Martin Luther King had been assassinated, the same fate that would befall Democratic presidential candidate Robert Kennedy a month later. As the Beatles sang about a "Revolution," scores of mostly young people swarmed Washington and college campuses to protest Vietnam and call for social justice. Alongside this sea change was a burgeoning eco-awareness that would lead to the first Earth Day two years later.

CHRISTENED—Councilman Louis R. Nowell breaks champagne bottle over bulldozer to symbolize start of work on the Foothill Freeway. Site is at Foothill and Glenoaks Blvds City, county and state officials participated. Among spectators were, from left, T. R. Lammers, David Negri and Ted Milliner, officials.

Louis Nowell christening the Foothill Freeway. The councilman's antics were catnip for L.A. newspapers.

As 1968 drew to a close, Frank Lanterman carried this torch of environmental activism into a transportation planning committee just before Christmas. The assemblyman was not some johnny-come-lately tree-hugger; he had the track record to back it up. The grandson of the co-founder of La Cañada, Lanterman had deep roots in the area. One of his first jobs was as the organist at the Alex Theatre in Glendale. During his many terms as a legislator, he sponsored several bills that addressed noise and air pollution and paved the way for anti-smog laws in California, such as the establishment of motor vehicle emission standards. Unfortunately, there was nothing Frank could do about the bulldozers about to carve up his beloved Crescenta Valley. His only weapon now was his words, and he used them to counsel committee members about what was at stake.

"Somewhere [you] have forgotten the people's benefits," he intoned. "Now they will have the death-dealing fumes of

automobiles." The Foot-
hill Freeway didn't just
place "truck benefits"
over people; it was an
indictment of the Amer-
ican way of life. In Eu-
rope, he explained, a
serene, clean-air valley
like La Cañada would
be protected. His views
were backed up by a guest
speaker named Frank J.
Tysen, a researcher with

The Foothill Freeway running northwesterly through the Verdugo Mountains and Crescenta Valley. Visible in the upper frame, the whitish outline of Big Tujunga Creek and Tujunga Wash, which feed the Hansen Dam flood control basin just west of the freeway.

the Institute of Urban Ecology. "The blatant disregard for social environmental values which has characterized freeway planning has caused a nationwide revolt," he warned. People were rising up because "we have failed badly with our freeway planning or rather transportation planning" due to the fact that "freeways have been planned by the wrong people—highway engineers." He called for more attention to preserving the natural surround-ings, a throwback to the days when the speedy Arroyo Seco Park-way accommodated scenic splendor. James A. Hayes (R-Long Beach), the vice-chairman of the committee, dismissed Tysen's impassioned plea as "half-informed and somewhat arrogant."

The activists may have been losing the short-term battles, but they were winning the war. Within weeks, they would soon get an assist from Mother Nature herself.

From January 18 to 27, 1969, Los Angeles was pummeled with nine straight days of rain—the worst deluge since the deadly floods of 1938 that led to $100 million worth of improvements to its flood control network. Earlier that month, the *L.A. Times* had cited the infallibility of this network. But it proved quite fallible. A bridge at Big Tujunga Wash got washed away, stranding 200 homeowners. And they were the lucky ones. A dozen people were

Epic failure: A neighborhood in the floodplain during a flood control breach.

buried alive by mudslides plowing into homes in the foothills. Fifty-two died in rain-related accidents. Another nineteen drowned in storm channels or cars that had veered into riverbeds, unable to get out before the rising waters swept them away. At a flooded compound housing 300 animal "actors," trainers had no choice but to release herds of deer into the wild, though they managed to save the feline star of the MGM film *Clarence, the Cross-Eyed Lion*. With damages north of $30 million, President Richard Nixon declared a disaster zone.

Just as the city dried out, another front barreled in. By mid-February, twenty-five inches of rain had drenched Los Angeles, the heaviest season in eighty-four years. Again, Big Tujunga Wash failed. A torrent of water 150 yards wide and 40 feet deep, racing along at thirty miles per hour, overflowed its channel walls, whisking seven houses downstream. "NORTHEAST VALLEY AREA 'DESOLATED,'" the *Valley Times* shrieked. Work was paused on the 5.6-mile stretch of freeway where it passed through the Big Tujunga alluvial plain near Hansen Dam. Engineers needed to assess the damage and consult with the Flood Control District on next steps. Like it or not, they had plenty of time. In March of 1970, Congress deliberated on a new highway bill and froze $32.4 million in funds for California, much of it earmarked for the Foothill Freeway. Construction wouldn't start again for six months.

Proving the "when it rains, it pours" truism, the Foothill Freeway suffered another setback when the 6.6-magnitude Sylmar Earthquake hit in early 1971. Mere months from completion,

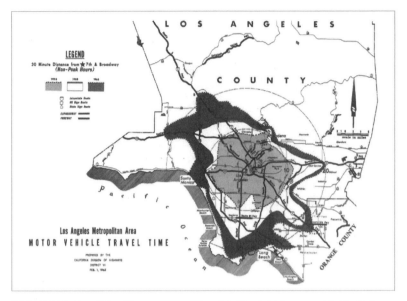

"Motor Vehicle Travel Time" from a 1965 edition of *California Highways and Public Works*. Each radial shading represents how long it takes to drive 30 minutes from 7th Street and Broadway during non-peak hours. The gray field represents the year 1953; the white ring is 1958; the dark gray outer-ring is 1965, by which time freeways enabled motorists to make it from Downtown to Long Beach, Santa Monica, or Sylmar within a half-hour.

the $15 million Foothill-Golden State Interchange straddled the epicenter; several of its bridges buckled and toppled, including the western connector of the 210 West to the 5 South that killed a motorist and his passenger. Access to the freeway's western extremity was set back three years.

The following year, another Foothill Freeway span fell—this one, eerily, near Devil's Gate Dam. Construction workers pouring concrete were crushed when a sixty-foot section of the half-built bridge collapsed into the gorge. Six dead bodies were pulled from the piles of rubble, two of them entombed in the quick-dry cement. Twenty-one more were injured. Initial reports were that another earthquake had struck, but the fault lay with a lax contractor. "The lousiest construction firm I ever had to deal with," a state safety engineer later testified. It would be almost two years before *that* bridge was fully rebuilt.

In the midst of these tragedies, a modest stretch between La

Cañada and Pasadena opened to traffic. But an aura of freeway fatigue hung in the air at its anachronistic dedication, which advertised "pretty girls in island costumes" swaying to Hawaiian music (don't ask). Even the master of ceremonies—usually an unabashed booster type—couldn't mask his cynicism. "I thought the Foothill Freeway . . . was the most ridiculous waste of public funds in the nation," pronounced Warren Dorn from the Board of Supervisors. "But that is behind us now. You win some. You lose some." As the ribbon was cut to a smattering of applause, three young men on bicycles crashed the party by holding up signs that read, "Hiss! Hiss!"

FRAYED NERVES AND THE MISSING LINK

Back when Councilman Louis R. Nowell was photographed gleefully smashing a champagne bottle at the groundbreaking for the 210's western gateway, the ardent freeway advocate had every reason to believe he would be presiding over a completion party in 1972. But now the year had arrived, and the roadway still had a ways to go.

The intervening four years had not been kind. Mudslides, floods, frozen funds, an earthquake, and bridge collapses were only the tip of the iceberg. The recent passages of the National Environmental Policy Act and the California Environmental Quality Act called for new construction to be considered against its natural surroundings. This had immediate implications for the Foothill Freeway—in particular, the 5.6-mile missing link between Van Nuys Boulevard and just east of Sunland Boulevard. After the storm damages of 1969, Nowell endorsed a proposal to construct a levee in Big Tujunga Wash to divert runoff to a concrete channel, thereby protecting the freeway and its vicinity from more floods. Lawmakers were on the verge of taking the project to bid when the Federal Highway Administration, citing the new laws, suddenly told them to stop. No work could commence until the

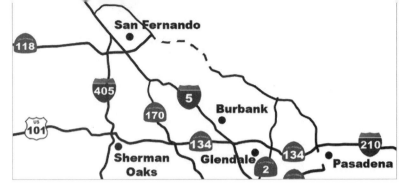

The 5.6-mile "missing link" through the Tujunga watershed—represented by the broken line—was the result of a protracted battle between conservationists and freeway lobbyists.

state submitted an Environmental Impact Report.

Nowell was livid. Things used to be so much easier when you could just ram projects through with local politicos, save for a few winking quid pro quos in smoky back rooms and bars. With the feds now in their face, work on the gap was stalled until late 1974—one year for Caltrans to complete an EIR, and another year for Washington to review and approve it. Only then would freeway engineers know how long, how wide, and how high a causeway or levee across Tujunga Wash should be.

Nowell blasted the namby-pamby ecofreaks who had created this whole mess. For the last seven years, he had been trying to convince them to purchase 600 acres around the wash to preserve its natural habitat, which could have obviated the need for a levee. "It has always been a mystery to me why many of the so-called environmentalist groups have utterly failed to support the purchase of this area which they claim is so vital," he said. James Stanley, president of the City Traffic Commission, also took a shot at this group. A study of soil in the contested area "required getting botanical names of every plant eighteen inches high," he huffed. "Now, they want another study of plants twelve inches high."

Called out on their credibility, dozens of environmentalists— many of them members of the Coalition to Save the Big Tujunga Wash—showed up at the next town hall meeting in February of 1973 to upbraid the councilman. A dike would eviscerate one of the last remaining natural streambeds in the L.A. area. It would cause erosion and clog up Hansen Dam with sediment (this exact problem occurred at Devil's Gate Dam, where engineers embarked on a project in 2021 to eliminate detritus from the San Gabriel Mountains). Presaging the anti-dam movement that took off in the 1980s, the coalition called for building strong earthen banks along the stream, keeping the flood plain wide, and letting nature run its course. They also pointed out that the dry conditions from a dam would invariably lead to gravel mining downstream. Then, they leveled an accusation of their own: Maybe that had been Nowell's plan all along.

Nowell denied soliciting funds from gravel interests, but it certainly wasn't the first or last time he was accused of bribery. In his fourteen years as a councilman in the San Fernando Valley, Louis was fined for numerous violations—many tied to pro-development schemes. Like the time he rented a boat slip in Marina del Rey from a real estate mogul. In return for a 50 percent discount, Nowell voted to approve the developer's controversial condominium project along the beach. Or his all-expenses-paid trip to Puerto Vallarta, Mexico, bankrolled by a billboard company. When he returned to L.A., he voted against a city measure to control billboards. As always, Nowell was unapologetic and pooh-poohed any wrongdoing. "Next time I'd insist that they send me to South America and make a big trip out of it instead of a little hop to Puerto Vallarta," he said defiantly.

One thing about Nowell was on the level: he was the most developer-friendly member of the city council, quick to vote for zone changes and carving up mountains to make way for roadways and residences. When Nowell endorsed a plan to dynamite Franklin Canyon for a four-lane highway that would trample Beverly Drive, Beverly Crest couple Jerry and Debbie Decter were so appalled that they made it their life's mission to bring the councilman down. "The deeper we dug, the more we realized Nowell was up to his eyeballs in mass development and in freeways," Jerry said.

Nowell's response to people like the Decters: "You can have your friends sue me. That doesn't trouble me."

As highway officials labored over the Environmental Impact Report, they were making hay elsewhere on the 210. Just south of the Rose Bowl—where the Foothill Freeway, Ventura Freeway, and an unfinished spur of the Long Beach Freeway came together—the interchange was homing in on completion. And after a three-year delay, engineers were readying the resurrection of the I-210/I-5 Interchange that had collapsed in the Sylmar Quake. Also, much of the freeway between Pasadena and Pomona was finished.

In February of 1974, Caltrans finally completed the EIR regarding the 5.6-mile gap—six months later than anticipated. The public was invited to peruse copies of the 250-page report at Valley libraries. After residents weighed in, the city council and other agencies offered their input. State officials then sent it off to Washington and held their breath.

In the meantime, the freeway follies continued over the summer. Two more bureaucracies threw up roadblocks to the Tujunga Wash gap that almost seemed designed to further poke Nowell in the eye. The California Air Resources Board—charged with implementing new Environmental Protection Agency standards—concluded that extra traffic from the 210 would generate unacceptably high levels of contaminants. The county Air Pollution Control District warned that auto-related pollutants would form a "smog trap" over Pacoima, enough to curtail physical activity at three schools within 200 yards of the freeway. Pacoima was always a raw spot for Nowell, who accused its large Black community of living off welfare programs from cradle to the grave. "Are they going to do it for themselves or leave it up to big daddy and big mommy?" he asked.

Nowell's policies nearly led to a fistfight with a Black councilman named Dave Cunningham. "You are the greatest racist in this world!" Cunningham shouted. Nowell's personal beliefs aside, another extended holdup was not a good look for the self-professed friend of free enterprise. Nowell immediately fired off letters to

The Glendora Curve, facing east, marking the spot where the 210 becomes a State Route at its junction with SR-57. Prior to 2002, SR-57 was still part of I-210, when the interstate's terminus was the Kellogg Interchange in Pomona. The Foothill Freeway now ends in Redlands.

state and federal officials, urging them to institute "emergency" construction. His request fell on deaf ears.

By 1976, the final pieces of the Foothill Freeway puzzle were coming together so lethargically, even the *L.A. Times* couldn't resist sarcasm when announcing the 210's final leg through Pasadena: "FOOTHILL LINK OPENS TUESDAY, WANNA BET?" read their March 11 headline. Amazingly, the Pasadena section opened as planned. Now Nowell *really* had something to seethe about. Every single mile of the 48.5-mile Foothill Freeway was now complete except for the 5.6-mile gap in his district. In a year that saw the release of the movie *Network*, Louis Nowell went full-on Howard Beale—the bitter network anchor who says anything on his mind as he slowly loses it. That September, presiding over a ribbon-cutting for a 3.3-mile section of the Simi Valley Freeway that ran through his district, Nowell remarked that he was dedicating "actually nothing. This freeway section goes nowhere." Nowell was referring not just to his own white whale, but to the unfinished Simi Valley Freeway, also mired in environmental roadblocks. "It's made virtually unusable by the gap to the west and is stymied to the east by a missing link of the Foothill Freeway crossing

One consequence of the Foothill Freeway's dormant state through the northern Valley: mountains of garbage in idle construction zones, making a mockery of the state's "No Dumping" signs.

Big Tujunga Wash . . ." He ranted about all the "selfish people" holding up millions of dollars to finish the 210, and jeered ill-advised transit ideas favoring "choo-choos" over cars. Commence awkward slow clapping.

Nowell's frustration reflected much of his district. James Moran, a longtime resident, complained that environmentalists only cared about "the flowers and the birds and bees." Chet Mc-Donalds, president of the Sunland-Tujunga Chamber of Commerce, observed that five years of environmental fights over the missing link had led to economic stagnation. "Developers and investors are turning away from this community because they are not sure a freeway will ever be built," he said. It was certainly beginning to look that way, especially after it was revealed that the state was preparing two more Environmental Impact Reports related to the freeway's path. One involved an underground water source at an abandoned gravel pit that the Fish and Game Commission wanted to make sure didn't harbor rare fish (it did; they were relocated). The other centered on whether a certain patch of land was a sacred Indian burial ground (it wasn't). For those keeping score at home, the state was juggling at least five different EIRs for one 5.6-mile ribbon of freeway.

It was all finally too much for Louis Nowell. At the end of 1976, he announced he would not be seeking another term. The travails of the Foothill Freeway, however, were far from the sole reason. As promised, Frank and Debbie Decter continued to hound the councilman, using their own resources and enlisting L.A.'s district attorney to expose his

OFF-RAMP ↗

1976 was a watershed year for the conservation movement. Greenpeace drew worldwide attention when its members risked their lives in boats by blocking harpoons from whaling ships. More locally, vigilantes in Pasadena fought back against Caltrans's plans to tear down oak trees in the path of the proposed 710 Freeway extension. A man threatened to chain himself to a tree in protest, and several women posted signs blasting Caltrans and warning of a "Tree Killing Area."

FREEWAYTOPIA

fraudulent activities and drive him out of office. In the words of Council President John S. Gibson Jr., Nowell became "a tense, distrustful, and people-hating man. He began thinking everyone was against him and that he couldn't trust anyone." In his final farewell to the city council, his voice cracked as he broke down in tears. Readers of the *L.A. Times* were unmoved. "I say three cheers for the resignation of Louis Nowell!" wrote Pasadena resident Jerry Stone, speaking for many who condemned his arrogance and shiftiness.

One subset that loved the dormant construction of the 1970s were street racers, who always found a way to access forbidden freeways to open up their muscle cars. The Tujunga section was a favorite for years. Speaking of which, another group that took advantage of the downtime were Hollywood producers. *Corvette Summer, Death Race 2000, The Gumball Rally,* and *Cannonball* were among the schlocky hot rod movies that filmed on the unfinished Foothill, Glendale, and Simi Valley Freeways. But nobody got more mileage out of the situation than the TV show *CHiPs.* The popular NBC action drama followed the exploits of two strapping California Highway Patrol officers in tight khakis and black leather boots—Larry Wilcox's Jon and Erik Estrada's "Ponch." As the CHP's domain, L.A.'s freeways figured prominently in the series—so much so, Columbia Pictures Television rented the deserted portion of I-210 from 1976 to 1980. The upshot: Lots of gratuitous chase scenes and pile-ups, not to mention sexy shots of Ponch and Jon cruising around on their Kawasakis—all put to a disco music bed, of course. But it wasn't all fun and games in the fast lane. Our heroes

A bromance for the ages: *CHiPs* partners "Ponch" (Erik Estrada) and Jon (Larry Wilcox).

OFF-RAMP ↗

There would be no *CHiPs* if the California Highway Patrol hadn't taken over control of the Los Angeles freeway system on October 2, 1969. Prior to that, L.A.'s freeways were patrolled in a piecemeal fashion by 282 police personnel drawn from dozens of communities. Senator Randolph Collier was the first to propose CHP oversight of freeways in 1955. Fun fact: *CHiPs* headquarters on the TV show was portrayed by a real Highway Patrol office, located under the Santa Monica and Harbor Freeway Interchange.

A *real* California Highway Patrol car driving under the access ramp from the 110 South to the 10 East.

also stopped a runaway tractor-trailer, rescued kids from a cave under the freeway, and saved babies from botulism-tainted baby food that spilled off a cargo truck. Honestly, the Foothill Freeway factored into so many storylines, it deserved its own special Emmy (along with Erik Estrada's famous teeth).

The free ride for filmmakers ended on April 3, 1981. That's when work on the Foothill Freeway finally, improbably, concluded. While no one quite got what they wanted, the resulting détente was an ecological victory. Two bridges were built over the Tujunga Wash, high enough to allow free movement of wildlife, hikers, and equestrians in the channel's dirt bottom. There was no levee. High-walled earthen banks were embellished with natural vegetation, and contractors were required to follow a litany of strict environmental guidelines (such as trucking waste off-site) that would be a model for all future roadway construction. But the delays came at a hefty price. In 1971, the budget for the 5.6-mile patch was $16 million. Ten years later, it was completed for $42 million—the costliest road segment in California at that time. But for nature-lovers, it was money well spent.

At the dedication attended by Mayor Tom Bradley, the Verdugo Hills High School Marching Band performed alongside an antique car club and a color guard. Not present: retired Councilman Louis Nowell. He was chilling out on his yacht in the marina.

FROM FOOTNOTE TO FLASHPOINT

Although Interstate 210 was now finished, it would soon be "unfinished" by Caltrans's decision to relocate its eastern terminus. The agency announced that the leg from the "Glendora Curve" to the San Bernardino Freeway (I-10) would be decommissioned and renamed (it became State Route 57, or the Orange Freeway). The Foothill Freeway's eastern section would be rerouted along the former California 30 corridor at the base of the San Gabriel Mountains, ending in Redlands. This would change the overall length of the freeway from 48.5 miles to 86 miles, built in two phases at any eye-popping $700 million. In the early 1990s, construction on phase one commenced from Glendora to San Bernardino. None of this registered as news to the average Angeleno, whose familiarity with the Foothill Freeway—a fringe freeway for most—was now associated with the savage beating of African American motorist Rodney King.

In the early morning of March 3, 1991, King was heading west on the 210 after a night of drinking with two friends. All three were in the car when CHP officers Tim and Melanie Singer—a husband-and-wife team—spotted King's Hyundai driving at high speeds. They began a pursuit, which reached 117 miles per hour along the wide-open freeway. King exited near the Hansen Dam Recreation Area. As the chase continued on surface streets in Lake View Terrace, the Singers called the Los Angeles Police Department's Foothill Division for backup. LAPD units forced King to a stop on Foothill Boulevard and he was ordered out of

A screen grab of officers beating motorist Rodney King, the flashpoint for the civil unrest that would follow in 1992. Plumber George Holliday shot the footage with a Sony Video8 camera from his Lake View Terrace balcony.

the car. After resisting handcuffs, King was tased and brought to the ground.

What happened next is fixed in the consciousness of many Americans: a Möbius strip of endless kicks and baton strikes to his head and body, all captured on a camcorder by eyewitness George Holliday—considered by many to be the very first "viral video." Four officers were acquitted of assault at the 1992 criminal trial, leading to widespread civic upheaval that killed 63 people and injured 2,300 over six days. After a troubled life, King would later die from an accidental drowning in 2012.

Transcripts from CHP and LAPD communications showed strikingly different approaches on how to resolve the unfolding situation. As the Singers pursued King on the freeway, they reportedly intoned over their bullhorn, "Pull over to the right. We won't hurt you." When King emerged from his car, Melanie observed him to be jovial and dancing—a sloppy drunk. When she moved in to arrest King, she was shunted aside by LAPD Sergeant

Stacey Koon, who told her and her husband, "No, no. Get back. We'll handle." The Singers and a third CHP officer, Gabriel Aid, were relegated to keeping an eye on King's two passengers. Tim Singer would later say of King's beating, "It reminded me of a monster movie, where the monster gets shot and is still coming at you," making it clear that the "monster" in this movie were the L.A. cops' fifty-plus blows to King.

Rodney King.

All three CHP officers considered King to be subdued and were so alarmed by the unwarranted blows, they contemplated getting the officers' names from their ID tags. Even at the hospital afterwards, Melanie assumed that the CHP would retain jurisdiction of the case since it began on the freeway. But when she started her paperwork, Koon came over to tell her his department would handle it from here. Koon and his partners tried to intimidate King at the hospital as he gave his statements. It was revealed the cops had made racist jokes, and Mayor Bradley called out the department for bigotry and its "dangerous trend of racially motivated incidents." King won a civil suit three years later.

Rodney King, of course, made the choice to lead authorities on a high-speed chase—he was on parole already and didn't want to be arrested—and he was initially uncooperative when pulled over. But there exists an alternate universe where his arrest for intoxication and speeding on the Foothill Freeway was handled only by the agency that enforces the freeways—the California Highway Patrol. Perhaps history would have been vastly different if Gabriel Aid and Melanie and Tim Singer had arrested Rodney King instead of fifteen LAPD officers.

THE LAST FREEWAY?

The new eastern extension of the Foothill Freeway to Redlands—where it rejoins Interstate 10—opened to traffic on July 23, 2007. Paid for by the state, these forty miles fall under the California Freeway and Expressway System and are thus designated SR-210 instead of I-210. From start to finish, it took fifty-two years to complete the Foothill Freeway—the longest timeframe in the Southland's freeway system (though to be fair, the initial route to Pomona took only half that).

Because it began in the bullish era of freeway-building and ended during the anti-freeway movement, the Foothill Freeway was a living barometer of public sentiment over the course of its progress—or lack thereof. It's no coincidence that the ten-year delay of the 5.6-mile patch happened during the 1970s, when the expense and hassle of freeways in an increasingly smoggy, car-centric city was compounded by community blowback. It was the decade when virtually every last freeway was purged from the planning dockets—among them the Beverly Hills, Reseda, and

OFF-RAMP ↗

Because the designation of interstates falls under the jurisdiction of the American Association of State Highway and Transportation Officials, Caltrans has petitioned AASHTO to re-sign the entire Foothill Freeway as I-210. Highway geeks are watching with bated breath . . . perhaps it will have been approved by the time you read this.

MTA's L Line (formerly the Gold Line) runs along the Foothill Freeway's median from Pasadena to Arcadia. Brought online in 2003, its right-of-way through the San Gabriel Valley dates back to the 1880s.

OFF-RAMP ↗

Though its outlook was bleak in 1980, the Century Freeway (I-105) overcame environmental and community hurdles and managed to open in 1993. And of course, there was the freeway segment that refused to die—the 4.5-mile link that would make the Long Beach Freeway (I-710) continuous from Long Beach to Pasadena. After six decades on the books, a legislative stake was finally driven through its heart in 2018.

Pacific Coast Freeways. In the 1950s, the Division of Highways planned 1,557 miles of Southland freeways. By 1980, only one-third had been laid down. It was believed that the Foothill Freeway represented the last hurrah. As we'll later see, reports of the death of freeways were greatly exaggerated.

Again, a common misperception is that the term "freeway" got its name because said roadways were free of tolls. Others interpret freeways as a derivation of freedom—freedom to move goods, freedom to roam, freedom to get stuck in traffic. For the Louis Nowells of the world, the Foothill Freeway was a continuum of this distinctly American conceit. But it turns out freedom wasn't everything it was cracked up to be, or at least it meant different things to different people. As we'll see later, the Foothill Freeway provided a progressive roadmap for the Century Freeway. The scales of justice were now equally weighted, with freeway fighters on a more even playing field against freedom fighters.

Saddled against the San Gabriel Mountains, the Foothill Freeway provided a front-row seat for the 2009 Station Fire, at the time the largest wildfire in Los Angeles County history.

Chapter 6

THE VENTURA FREEWAY
U.S. Highway 101 / State Highway 134
(1955–1974)

The Ventura Freeway is so ancient, it predates Los Angeles.

Okay, so maybe "freeway" isn't the right term. But it *is* part of U.S. Highway 101, which was built in the footprint of El Camino Real—one of the oldest routes in California. Though the 101 extends over the length of California, the Ventura Freeway portion starts at the Santa Barbara/Ventura County line. From there it runs south to Oxnard, then makes for the City of Angels. After sixty-five miles, U.S. 101 becomes the Hollywood Freeway, whereas the Ventura Freeway continues eastward as State Route 134 to Pasadena. You could say that the road has a split personality. This wouldn't be far off, though its path may be more analogous to the stages of life.

From Ventura through the Conejo Valley, the freeway is a carefree melody of coastal breezes and strawberry fields in Camarillo. This is the roadway of America's "Ventura Highway," conjuring up youthful good vibes. Once it enters Los Angeles County—the focus of this chapter—the soulful bro enters middle age, just another working stiff cursing in traffic. Then, passing the 5 Freeway, the freeway mellows out again, retiring in picturesque Pasadena alongside the comforting confines of the old Colorado Street Bridge.

Most Angelenos, though, have to dig deep to love the Ventura Freeway. Almost everyone has their own war stories involving its junction with the San Diego Freeway. If freeways are L.A.'s cathedrals, then this meat-grinder interchange is the Church of Perpetual Congestion. Burdened by a flawed design and endless

Agriculture remained a dominant industry in the San Fernando Valley through the 1930s. A local paper profiled the Wood family and their berry fields in 1935.

suburban sprawl, it is a daily reminder that some things in our lives outlive their usefulness. But we return to them anyway, time and again, not because we want to, but because it's still the best option out there. After all, the Ventura Freeway is the only west-east thoroughfare that cuts across the entire San Fernando Valley. If acceptance is a form of surrender, Angelenos have long since made truce with the Ventura Freeway.

On December 2, 1953, representatives of the California Highway Commission called a special meeting about the Ventura Freeway at the State Building in Downtown Los Angeles. Most of the hundred attendees were property owners from communities like Woodland Hills, Tarzana, and Canoga Park. Commissioners had already announced the adoption of the Ventura Freeway through the southwestern San Fernando Valley. They called it Route A. It would mostly run alongside Ventura Boulevard, the Valley's main drag. Even though the route was a fait accompli, two factions squared off—one group wanted Route A to swing along

PENCER TRACY ESTATE ENCINO, CAL. X325

Celebrity estates like Spencer Tracy's, once hallmarks of the San Fernando Valley, continue to fascinate. The Valley Relics Museum has hosted multiple exhibits of "Old Hollywood Ranches of The SFV" in recent years.

train tracks farther north; another wanted to keep Route A the way it was, if only to eliminate any more uncertainty.

Then there was Harold Bayly.

Lording over an expansive agricultural empire in Woodland Hills, Bayly was on a two-month European sojourn when the commission announced the first stretch of the Ventura Freeway in L.A. County—10.6 miles from Calabasas to Sepulveda Boulevard at $20 million. When he showed up at the hearing, Bayly looked more like a Wall Street mogul than a countrified rancher in his three-piece suit and crisp pocket square. With him was a mini-entourage that included his jowly lawyer, T. B. Cosgrove, and a private civil engineer, Marshall Pond. They were here to do the devil's work, to strong-arm the state into considering the alternate Route A that would spare Harold's ranch from condemnation. Such was Bayly's influence that the state hastened an emergency parley to placate him.

Bayly's ranch was one of dozens that had dotted the West Valley since the 1920s. Many were owned by movie studios, which found the Valley's rambling landscape well-suited for Westerns.

Celebrities were also drawn to its slower lifestyle. Comic actor Edward Everett Horton owned an estate called Belly Acres, where F. Scott Fitzgerald rented a guest cottage in the late 1930s. Betty Grable had a ranch with her big-band leader husband, Harry James.

OFF-RAMP ↗

Edward Everett Horton lives on in Encino. A short street is named after the comedian off Burbank Boulevard, close to the address of his original ranch at 5521 Amestoy Avenue, which was taken out by the freeway.

Even Bayly's farmstead hosted dozens of open-air movie scenes, including Douglas Fairbanks's film debut. And of course, Edgar Rice Burroughs, the creator of Tarzan, bought up so many acres in the community of Runnymeade that the townsfolk voted to change its name to Tarzana.

But by the early 1950s, the insatiable, hungry caterpillar that was the freeway industry was gobbling up the fringes of L.A. County. Tucked away in the Santa Monica Mountains, Fox Ranch and Paramount Ranch lay far enough south to be unaffected by the Ventura Freeway. Horton, Grable, James, and Bayly drew the short straws—the original Route A was slated to slash through their domains. Only Bayly came to the meeting with his own engineering expert, who insisted the state could simply revise the route to go up and over his client's threatened ranch in the form of a cartological speed bump.

Then, in a twist right out of a courtroom drama, Bayly was upstaged by the sudden appearance of his neighbor—studio head Harry M. Warner of Warner Bros. Warner owned 1,000 acres of ranchland that would also be impacted by the freeway. Born Hirsz Mojżesz Wonsal, the make-good descendent of poor Polish Jews conveyed a cinematic tenor. "I have faith in the American principles—that's why there are commissions like this," he pronounced. "Don't worry about me and my property. Do what's best for the people. And if I lose all my property, thank God I'll have lost it in America!" Warner's patriotic speech was met with hearty applause, which seemed to warm the cockles of his old heart. (It was later revealed the state offered him $7 million for the portion of his ranch needed for the freeway, which would warm anyone's cockles.)

The meeting adjourned, but Warner's capitulation was revealing. Two weeks later, a Southland newspaper declared: "STATE STANDS BY ROUTE FOR VENTURA FREEWAY." It was a quick, knockout blow to Bayly, and a clear message to the sons and daughters of Hollywood: not even the most powerful forces in Los Angeles could stop the juggernaut that was the California Highway Commission.

CONCRETE PARADISE

As Los Angeles's unrelenting growth pushed westward, the commission began to bolster its highway infrastructure. In Santa Barbara, it spent $3.5 million to convert 2.2 miles of U.S. Highway 101 to freeway standards in a section prone to accidents. Freeway conversions also took place between Newbury Park and Calabasas. By 1955, the state had already completed over twenty miles of freeway from Santa Barbara to Los Angeles County. (Like the gradual conversion of the Ridge Route on Interstate 5, however, some stretches were technically four-lane expressways; "freeway" was often a catchall term for a divided highway. These segments would later be upgraded to freeway thresholds.)

The Ventura Freeway's southern destination was the Hollywood Freeway's then-terminus near Vineland Avenue. For this reason, it was often referred to in the 1950s as the Hollywood Freeway Extension (different from the eventual Hollywood Freeway extension that became SR-170). It was also called the Riverside Freeway (a nod to the street alongside the Los Angeles River), the Ventura Freeway (referring to its control city, even though the 101 continues northward as a freeway), and the Riverside-Ventura Freeway. All told, the journey from Ventura to

This 1956 map reflects the patchwork progress of the Ventura Freeway, then called the Riverside Freeway east of the Hollywood Freeway and stopping at the Golden State Freeway.

the Hollywood Freeway would comprise sixty-five miles. State Route 134—its elongation to Pasadena—was still just a glimmer in engineers' eyes.

A freeway to the West Valley couldn't come soon enough. In the decade after World War II, the region's population doubled and land prices tripled. From 1950 to 1960, the entire San Fernando Valley added 430,000 residents—an increase of 138 percent. To put the Valley's migration in perspective, in 1950, the region accounted for 16 percent of Los Angeles's total inhabitants. By 1960, that figure had risen to 30 percent. During that same time, population percentages stayed virtually the same in the Westside and Harbor districts, and dropped from 68 percent to 52 percent in Central L.A.

As the state's Right-of-Way Department was busting up ranches, it was also buying out homes. Owing to lack of density, agents moved or destroyed about twenty dwellings per mile, a fairly low ratio compared to other freeway projects. By February of 1956, almost all land rights from Calabasas to Sepulveda Boulevard had been settled, enough for engineers to start construction on two bridges carrying the Ventura Freeway over Topanga Canyon and Ventura Boulevards in Woodland Hills.

While the freeway's routing through the West Valley was locked by the mid-'50s, east of Sepulveda was still fluid, thanks to a new development: instead of ending at the Hollywood Freeway, the Ventura Freeway would now sprout eastward as SR-134. Suddenly, Burbank, Toluca Lake, Studio City, and North Hollywood all had a stake in the action. Public hearings were predictably raucous and embattled. Just as the Highway Commission had devised a Route A for West Valley meetings, this time their planned alignment was labeled Route 1, which largely shadowed

OFF-RAMP ↗

On July 8, 1959, Governor Edmund "Pat" Brown signed a bill officially designating Highway 101 as El Camino Real from San Francisco to the Mexican Border. As tokens of its past life, replicas of Mission bells were placed along the 101 Freeway. The bells disappeared over the years, but started showing up again in the 2000s, thanks to a businessman named John Kolstad. After purchasing the entire inventory of the California Bell Company, the original makers of the bells, he convinced Caltrans to install them throughout much of the El Camino Real's old route.

A Mission bell along El Camino Real (U.S. 101).

Riverside Drive (briefly resuscitating the Riverside Freeway name for this 5.3-mile stretch). Many homeowners from Toluca Lake and Burbank hated this route. It would require the removal of 212 homes, 285 resi-

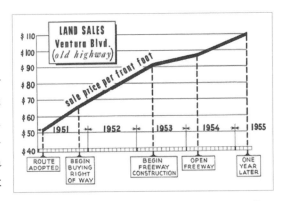

When the adoption of the Ventura Freeway was announced in 1950, it promised an influx of home buyers, businesses, and consumers. Sales prices per square foot of vacant land more than doubled along Ventura Boulevard within five years.

dential units, and 28 business establishments. This aggrieved group preferred Route 2—also called the "River Route," based on its proximity to the L.A. River—which would dogleg Route 1 for three miles between Universal Studios and the Lakeside Country Club. Due to fewer rights-of-way issues, the River Route would also be $8 million cheaper to build—certainly not chump change in those days.

After the debate over Route 1 and Route 2 dragged on, Paul Harding, state highway engineer for the L.A. district, urged everyone to close ranks on Route 1 and keep their eyes on the prize. This was the period when Congress was on the verge of passing the Interstate Act, which would crack open the federal piggyback for interstate freeways like the Golden State, Harbor, and future San Diego. Once that happened, Harding promised, "state funds allocated for those freeways could be diverted to the Riverside-Ventura Freeway and its construction could be expedited." Councilman Everett Burkhalter resented Harding's sword of Damocles tactics. He felt it was a ploy to get everyone to accept a route that many of his constituents disfavored.

As expected, the state adopted Route 1, which became the present-day path of the 134 between the 101 and 5 Freeways.

Burkhalter lambasted the state for its "arbitrary and egotistical attitude" and accused it of bowing to "political pressure" applied by Universal Studios and the Lakeside Country Club to keep the freeway out of their backyards. (Although, if you ask me, I would have loved to see a "freeway road rage" attraction on the Universal lot.)

Despite the threat of litigation, groundbreaking proceeded on September 5, 1956, for the future "Hollywood Split" interchange between the 134, 101, and 170 routes. And Harding proved to be a man of his word. In early 1957, the completion date for the Ventura Freeway was moved up two years due to the release of state funds that had previously been allocated for the Golden State Freeway, which was going through its own growing pains. Thanks to a heavily rural pathway and minimal right-of-way issues, construction costs to finalize the freeway from Ventura to the Hollywood Split were pegged at a reasonable $83 million.

Despite the Valley spats, Los Angeles's love affair with freeways was still in full bloom, with the Ventura Freeway a beneficiary of the heady times. On April 5, 1960, former Governor Goodwin J. Knight presided over dual dedications that

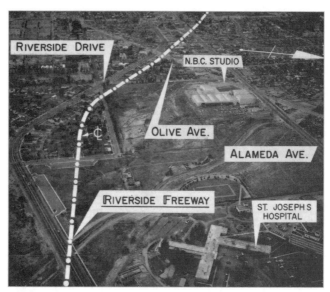

The 4.8-mile "Riverside Freeway" routing of the Ventura Freeway. Riverside Drive, retired as the main west-east arterial once the freeway opened, saw traffic plummet from 17,000 vehicles per day to 4,500.

bracketed the freeway's passage from Woodland Hills to Laurel Canyon Boulevard. Fifty antique cars rumbled down the pavement, though six of them conked out on the shoulder as a parade blazed past them. In lieu of a ribbon-cutting—and as a nod to the Valley's defense industry—Governor Pat Brown "pushed a button which raised a Nike missile into firing position . . . thus symbolically launching traffic."

OFF-RAMP ↗

Highway officials settled on the name "Ventura Freeway" for the multi-monikered roadway in May of 1955. At that point, the name only applied to the seventy-five-mile stretch from Ventura to the Golden State Freeway. However, after the extension to the Foothill Freeway was announced, everyone started calling the whole thing the Ventura Freeway, which the state officially adopted on February 5, 1957.

"VENTURA FREEWAY ACCLAIMED BY ALL," trumpeted the *Valley Times* the next day. The paper naively predicted that, for West Valley commuters taking the 101 to Downtown, traffic on Ventura Boulevard would become a thing of the past (while conveniently ignoring the impact all those cars would have on the 101). The article conducted experiments showing that the Ventura Freeway was safer than surface streets and cut travel times in half. Meanwhile, population centers proliferated in Ventura County—Thousand Oaks, Camarillo, Oxnard—the fastest-growing county in the U.S.

In the early '60s, the upstart freeway was still a novelty in the

"Gimme a V! Gimme an E!..." A Valley high school band dedicates a new segment of the Ventura Freeway in 1962.

Is that an unsafe lane change? Incomplete freeways were magnets for joyriding, be it hot rods or even go-karts. This photo of unknown boys, looking southeast, would have been taken no later than April 5, 1960, when the freeway opened to Laurel Canyon Boulevard (visible on the sign).

valleys of Conejo and San Fernando, and one can sense the fraying strands of innocence in what the media considered newsworthy: The "rodeo" that broke out when four cows jumped out of a cattle truck, tying up traffic for four hours. The young kids who hijacked an unfinished stretch for impromptu kite-flying. Eight-year-old Larry Gonzales of Van Nuys getting picked up by amused cops for pedaling his Schwinn along the freeway, resulting in a spanking by his mother. ("Just wait until your father gets home," she hissed.) And no freeway construction chronicle was ever complete without some lucky goofball safely crash-landing his rickety airplane. Running out of gas one night, pilot Ted Vaughn touched down on the 101 in Ventura. Slack-jawed onlookers watched him taxi through traffic to exit at Seaward Avenue, where he found a service station. But he couldn't fill up his plane. They didn't have the right gas.

OFF-RAMP ↗

Holy SigAlert! Adam West, star of the kitschy *Batman* TV series, lost control and flipped his car on the Ventura Freeway in February of 1969. He managed to escape serious injury. Newspapers referred to his vehicle as an "experimental" model. Could he have been test-driving a new Batmobile?

TIMES OF INTERCHANGE

During this frenetic freeway era, several routes were on target to intersect with the Ventura Freeway. In L.A. County alone, five planned interchanges were spread out in four- or five-mile increments, like vertebrae on a spine. To the west of the Hollywood Split, Oberg Bros. Construction won the bid for the 101's complicated exchange with the budding San Diego Freeway (the eventual I-405). Starting the job in late 1956, the firm somehow brought the ten-bridge maze online on July 10, 1958—at least six months ahead of schedule—kicking in bonuses on top of their $7,732,000 contract. "It's All Yours" read the caption under a picture in the *Van Nuys News* that week. The photo showed a shirt-sleeved Paul Metcalf of Oberg Bros. proudly directing state highway engineers toward the concrete leviathan with his index finger. The interchange opened the spigot for a stream of cars toward the western reaches of L.A. County.

Four years later, the Hollywood Split interchange came online. Its $6,478,000 price tag—$1.3 million cheaper than the Ventura-San Diego junction—reflected the fact that it lacks two transition ramps (an oddity explained in the Hollywood Freeway chapter). Governor Pat Brown flew back down for its dedication, a symbolically big deal at the time. Motorists on U.S. 101 (which morphs into I-5 south of Downtown) could now drive eighty miles from Calabasas to San Juan Capistrano without a single traffic light (the 405 wasn't done yet). The only hitch in the day's get-along: Brown's dull oversize scissors failed to cut the ceremonial ribbon. Councilman Lemoine Blanchard came to the rescue, pulling out nail clippers from his pocket. Brown used them to sever the final threads.

Farther east, the Ventura-Golden State and Ventura-Glendale freeway interchanges were completed in 1967 and 1969, respectively. Like the Hollywood Split, the Ventura-Golden State Interchange is another truncated job. There are no access roads linking the 134 East to the 5 North or the 5 South to the 134 West, forcing drivers onto surface streets to make those connections. As the $8.2 million interchange was being planned, funding shortfalls precluded it from being built as a full exchange. The state proceeded anyway, with project engineer George Dickey reasoning that "the work was authorized principally for the extension of the Ventura Freeway to the east to connect with the Foothill Freeway." Nonetheless, the state said the omissions were only temporary, and pledged to add those two critical ramps at a later date (decades later, motorists are still waiting). Not that anyone seemed to care when it opened. At the dedication on August 25, 1967, a calico goat on loan from the nearby Los Angeles Zoo stole the show. In lieu of the traditional ribbon-cutting, the beast nibbled through the ceremonial ribbon to mark the official "ribbon-chewing."

The final interchange, alluded to by Dickey, was the Ventura's junction with the Foothill Freeway (I-210) in Pasadena. This particular crossing traces its lineage to 1912. For almost half-a-century, the 150-foot-high Colorado Street Bridge was the main span across the Arroyo Seco between Eagle Rock and Pasadena. Known for its beautiful Beaux-Arts arches and globe lampposts, the registered historic landmark may just be the most romantic bridge in the Los Angeles vicinity. It's also the deadliest. From its inception to 1937, seventy-seven people jumped to their deaths, many perhaps drawn to the "Suicide Bridge's"

A modern view of the historic Colorado Street Bridge, spanning the Arroyo Seco.

This aerial view accentuates just how big an improvement the addition of the now-134 Freeway bridge was over its older sibling, which became so stressed by 1951, vehicles were banned during peak hours. The 134 Bridge is also known as the Pioneers Bridge, in honor of early transplants to Pasadena.

aching beauty as their final resting spot. (While not a complete deterrent, barriers have reduced suicides in recent years.) By the early 1950s, the two-lane bridge was sorely outdated. Chronic logjams made it clear that a new gateway was required to serve the growing needs of the western San Gabriel Valley.

Coordinating with the Pasadena City Council, the State Division of Highways commissioned a $6 million replacement bridge within a stone's throw of the old one. In the summer of 1953, the new bridge was completed. Initially six lanes wide with a divided center, its graceful supporting arches mimicked the original bridge, which remained as a frontage road. More importantly, the new span would serve as the centerpiece of a freeway flowing into the Foothill Freeway. No, not the Ventura. The Colorado Freeway.

As mentioned in the Golden State Freeway chapter, an appendage of the Colorado Freeway still exists off the 5 at the Colorado Street exit, but it peters out just past San Fernando Road. Four miles to the east, however, an eastern arm of the Colorado Freeway opened to traffic on June 25, 1954. Motorists entered from the intersection of Colorado Boulevard and Wiota Street and traveled one and a half short but speedy miles before reaching the

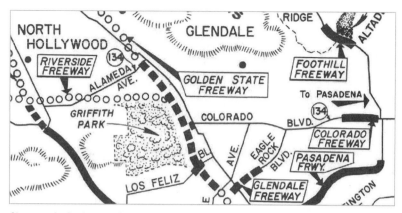

Close-up of a September 1956 map showing the then-Colorado Freeway. The extension of the Ventura Freeway (aka Riverside Freeway) past the Golden State Freeway had not been adopted yet, but note that Alameda Avenue and Colorado Boulevard already possessed its SR-134.

new freeway bridge at Linda Vista Avenue. Ultimately, the Colorado Freeway was merely a stand-in for a far bigger star. When the Ventura Freeway finally reached the Foothill Freeway in Pasadena, it subsumed its predecessor. A living ghost of the Colorado Freeway can be found in two strangely long access roads that used to serve as its entrance/exit: the Figueroa Street on-ramp to the Ventura Freeway East, and the Figueroa off-ramp from the Ventura Freeway West.

OFF-RAMP ↗

The Valleywide Committee on Streets and Highways had long lobbied state highway officials to extend the Ventura Freeway eastward to the Foothill Freeway. The committee also asked that the name "Colorado Freeway" be replaced with "Ventura Freeway." Although their request was granted, it would be years before "Colorado Freeway" dropped out of the local lexicon, so entrenched was the word "Colorado" in the Eagle Rock and Pasadena regions.

Like a giant ground sloth plodding eastward, it took twenty years for the Ventura Freeway to reach the new bridge, which had been freeway-ready since 1954. Meanwhile, while mid-Valley neighborhoods debated the freeway's alignment, East Valley residents debated the very notion of a freeway. In 1966, the State Highway Commission tagged 3,000 properties for acquisition

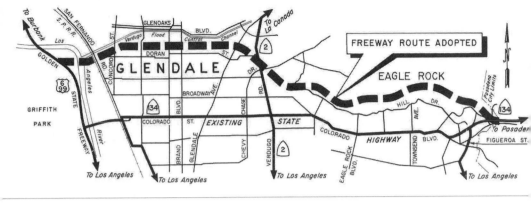

The Ventura Freeway's mapped-out extension through Glendale and Eagle Rock, May of 1961.

as engineers prepared to run the 134 under and over streets in Glendale. During the two-year construction effort, the sprigs of the environmental movement were starting to shoot in Eagle Rock—just as they were around this time for the nearby Foothill Freeway. Long known for their boho streak, Eagle Rock residents were galled by the state's decision to plow the Ventura Freeway through wild Eagle Rock Ridge—mere yards from the community's iconic rocky outcropping, which depicts a giant soaring eagle when shadows from the sun hit it just right.

That outrage was on full display for the Eagle Rock commemoration on August 18, 1971, when agitators jostled with advocates in front of the freeway's newly paved 2.3-mile segment. Members of Friends of the Earth, a student group from Occidental College, held up sloganeering placards that read "Millions for Freeways, Pennies for Clear Air," "Freeways Are Not the Answer," and "L.A.

Eagle Rock's signature rock, before and after the Ventura Freeway.

FREEWAYTOPIA

Needs Mass Transit Now." Councilman Arthur K. Snyder had their backs. The representative from Eagle Rock refused a seat on the dais alongside the mayors of Glendale and Pasadena and other dignitaries. When given a chance to speak, Snyder characterized the $9.8 million roadway between Harvey Drive and Figueroa Street as "an ecological disaster." Pointing to the San Rafael Hills, he said, "The tragedy is that it was placed through the most beautiful portion of the community."

Snyder was particularly critical of the way construction crews cut through the hills' ridges, exposing bald cross-sections of earth that looked like "a series of tombstones." Many of his constituents also complained about the eyesore, calling on state officials to expedite landscaping. (Snyder would later take their fight directly to Governor Ronald Reagan. The future president said that the route could not be beautified until funds allowed.) "Eagle Rock would be just as well off without this freeway," Snyder resolved. As for the day's ceremonial ribbon-cutting, the councilman huffed that it should be snapped by a garbage truck, since the only good thing about the freeway was that it would divert the trash trucks and their belching black diesel dust off Colorado Boulevard on their way to a local landfill.

The next-day's downbeat headline in the *L.A. Times*: "VENTURA FREEWAY CRITICIZED AT DEDICATION, CALLED AN ECOLOGICAL DISASTER." Quite a different tune from the previous decade's "VENTURA FREEWAY ACCLAIMED BY ALL."

Over in the West Valley, a

The Ventura Freeway has hosted its share of demonstrators over the years, including groups who blocked traffic in 2020 in support of Armenian causes. In 1981, these disabled men took to the freeway's shoulder to protest the suspension of bus service for wheelchair-bound passengers due to mechanical failures on the buses' lifts.

disaster of a different sort was playing out. In fact, it's a disaster that's never truly gone away: the dreaded Ventura-San Diego Freeway Interchange—arguably the worst-designed exchange in the entire Los Angeles freeway system. It would be easy to blame the Oberg Bros. Remember them? The contractors who wrapped up the

OFF-RAMP ↗

Another gripe by Eagle Rock residents was excessive noise from freeway traffic through the ridge. Throughout the '70s, the Highway Patrol deployed a Noise Abatement Team east of the Glendale Freeway Interchange. Cars exceeding eighty decibels (trucks were allowed ninety) were pulled over and cited by the noise police, with modified mufflers being the worst offenders.

complex project six months early? Within a few short years, that photo of Paul Metcalf handing off the freeway to state officials with the caption "It's All Yours" took on a whole new meaning, like that of a homeowner making off with a $7 million bag of cash after selling you a house on a bed of termites. By 1965, the Division of Highways had already admitted that "mistakes were made" when the interchange was constructed. The agency officially recorded the junction as a perpetual bottleneck. State engineers suggested twenty-three structural alterations to alleviate traffic. These construction outlays would well exceed the original cost—money the state didn't have—without addressing the root causes of the problem.

"It's all yours," indeed.

But context was also a factor here. The science of interchanges was still evolving in the early '50s. When the Oberg Bros. submitted their plans, they included narrow, single-lane connectors merging into *left-hand* lanes of traffic. If you're familiar with this interchange, no doubt you've experienced this yourself while switching from the 405 South to the 101 South, or the 405 North to the 101 North. This is a relic of the original design, an application already outdated by 1965. "I don't think we'd ever design a freeway again that brings traffic into the fast lanes," admitted

At 12 noon, congestion is already forming on the two far lanes exiting the 405 North for the 101 North, where they will merge with the Ventura Freeway's fast lanes—an engineering practice that has since been retired.

A. L. Muller from the Division of Highways, which signed off on the plans.

So what's wrong with traffic entering a freeway in the fast lane? Think about your average freeway. On- and off-ramps typically originate from the right-hand lanes. Seems sensible and orderly. But once you feed vehicles from another freeway into a new freeway's fast lane, these vehicles need to weave across three or four lanes of traffic to take the next exit. (Meanwhile, motorists *already* in the fast lane must slow down for merging traffic.) Next, throw in another stream of cars entering the same stretch of freeway from the slow lane. Now you've created double-weaving—cars entering from the fast and slow lanes, each looking to cross over to the other side. The 405's ramps for Ventura and Valley Vista Boulevards have also been continually readjusted for this problem. In short, the interchange is lousy with double-weaving, a contributing factor to chronic backups.

Though highway engineers had identified ways to improve the interchange by 1965, their plans coincided with a new era of budget austerity. Like the person who moved into a termite-infested house, the state was relegated to looking under the couch for spare change to buy ineffective bug spray from a hardware store. This led to years and years of piecemeal construction, which only exacerbated traffic. Frustratingly, when an improvement was made, it barely made an impact. Such was the case

Approaching the Ventura-San Diego Freeway Interchange from the south in September of 2020. A nearby brush fire prompted a SigAlert and led to miles of backed-up traffic. Is there a more apt metaphor for the troubled interchange than this photo?

when a second lane was added to the transition from the northbound San Diego to the northbound Ventura Freeway. "Congestion should not exist there now," said an engineer puzzlingly, perhaps overlooking the fact that, as of 1967, more than 80 million vehicles were already passing through the interchange every year. One reporter suggested that the state was telling Angelenos to practice mental jujitsu. "Drivers still caught up in that jam may find some relief by telling themselves the problem doesn't exist," she deadpanned.

As a new decade began, the *L.A. Times* devoted an entire section of their November 21, 1971, issue to the terminal problems of the interchange. Under an attention-grabbing headline—"SICK VALLEY INTERCHANGE MAY NEVER GET WELL"—the first paragraph offered a bleak prognosis:

> The afflictions of the Ventura-San Diego Freeway Interchange seem destined to last forever. It was born with arthritis, quickly developed clogged arteries and, despite massive surgery, will never be well. No one appears willing to pay the cost of a cure, even if one existed. Consequently, engineers have been forced to write construction prescriptions that hopefully ease the pain and concentrate, instead, on curbing the infection of increasing traffic.

Since blowing the whole thing up and starting over was not an option, the article laid out palliative measures proposed by the freeway's "surgeons," including ramp meters and freeway-widening. Other procedures were aimed at minimizing double-weaving—eliminating off- and on-ramps, or restricting access to them altogether. (An example of the latter would be the Haskell Avenue exit from the Ventura Freeway North. For motorists entering that freeway's fast lane from the northbound San Diego Freeway, lane barriers now prevent them from doing a 150-meter dash across five lanes of traffic to exit Haskell.) While many of these renovations were ultimately implemented, they've done nothing to lessen the quagmire that is the Ventura-San Diego Freeway Interchange. The *Times'* conclusion is no less relevant today than it was in 1971: "The thousands of drivers who creep through the interchange daily don't need to be told the system is sick—their feelings on the subject long ago became unprintable."

With no timetable to fix the headaches of the Ventura Freeway, politicians, journalists, citizens groups, and transit experts turned to more radical—some would say magical—solutions. Forty miles to the south, Disneyland's monorail system provided an intriguing template to cure L.A.'s traffic woes. Per the

This futuristic rendering of a monorail down the median of the Hollywood Freeway was just one of many sketches imagining the pods gliding noiselessly over L.A. freeways, including the Ventura.

Harbor Freeway chapter, monorails were explored for that freeway to ultimately connect to LAX. As early as 1963, the Alweg Corporation—the same company behind the monorail at Disneyland—offered to build a "highway in the sky" across the Valley. Its engineers envisioned forty-two miles of track at $123 million. Pointing out that the Wuppertal monorail in Germany had carried over one billion passengers since 1901, the *Ventura County Star-Free Press* called the transit system the natural choice to "relieve the monstrous traffic jams predicted for the '70s." Discussions with the Los Angeles County Board of Supervisors never advanced beyond the preliminary stage, but the monorail scenario stuck around and led to other elevated options.

At public forums in the mid-1960s, county commissioners introduced the idea of running a bus lane above the Ventura Freeway. At another meeting, representatives of the Canoga Park Civic Association proposed going full monty and double-decking the freeway to all traffic. They were roundly applauded by other Valleyites in the audience. Adding an upper level to freeways remained unpopular in greater Los Angeles, but these localized supporters of it had a ready-made bogeyman—the Whitnall Freeway. Before it was legislatively deleted, the Whitnall would mark an east-west course across the mid-Valley (straddling Roscoe Boulevard) before hooking southward—right through Canoga Park—on its way to Malibu. To the Canoga Park group, doubling the traffic capacity on the Ventura Freeway would erase the menace of the Whitnall Freeway through their backyards by rendering it redundant.

But dual-deck apostles encountered heavy resistance in the pricey foothills south of the Ventura Freeway. A towering thoroughfare would mar the view for residents in the Santa Monica Mountains and threaten property values. The Valleywide Streets and Highways Committee pointed to other double-deckers—the Alaskan Way Viaduct in Seattle, the Embarcadero Freeway in San Francisco—that had destroyed the character and quality of

The Valley's only realized monorail: The Skyrail at Busch Gardens in Van Nuys. The Anheuser-Busch park operated from 1966 to 1979.

life of their communities (both would eventually be torn down). "It would be ugly," said County Supervisor Warren Dorn. "The cost of building another layer on top of our freeways would be staggering, completely out of line."

And yet the stacked lane idea persisted into the early '70s. Dorn's colleague, Supervisor Frank G. Bonelli, observed, "Unless something is done soon, our freeways will not just be the world's longest parking lots during peak travel hours, but all day and night." His solution was to double-down on double-deckers. He envisioned second levels on the Golden State Freeway and the Pacific Coast Freeway before the latter even got beyond the blueprint stage. He called for the creation of private road authorities to finance and manage them—foreseeing the appearance of toll roads run by agencies like FasTrak by two decades.

Like the fast-talking monorail charlatan from *The Simpsons*, real-life Lyle Lanleys continued to pitch their schemes. Former aviation engineer Norman Greene vowed that his automated Greene Monorail would blow away the city's bus system with greater convenience at a fraction of the cost. The Rapid Transit District dismissed Greene's plan as "not practical for mass transportation." Harry Bernstein from Aerospace Corporation pitched a mini-monorail in the form of a Personal Transit System (PTS)—autonomous four-person capsules on elevated tracks above the freeway. The PTS was also DOA, essentially just a glorified people mover.

THE HIT PARADE

On September 25, 1974, Caltrans got one step closer to inaugurating the $21 million Ventura-Foothill-Long Beach Freeway Interchange. At ten o'clock that morning, workers removed the barriers between Orange Grove Boulevard and Fair Oaks Avenue. With the unceremonious opening of this 0.7-mile gap, traffic could now flow freely between the 134 and 210 Freeways. The Ventura Freeway—all 78.7 miles of it—was now complete (final touches on the interchange would take another year). As fate would have it, the freeway was already known by many people around the world before it was even finished.

Next to "Route 66," the Ventura Freeway may be the most famous non-mythical road memorialized in a pop song title. "Ventura Highway" was recorded by the band America and released on September 19, 1972. It was an instant classic, charting throughout 1972 and '73 in the United States, UK, Canada, Australia, and New Zealand. But wait—is it really about the Ventura Freeway? Well, remember that a freeway is technically a type of highway. According to songwriter Dewey Bunnell, he was in fact referring to the 101 Freeway through Ventura when he wrote the song, which calls on imagery from his youth in 1963 and '64. "That was my own misinformed self," he told *American Songwriter* journalist Paul Zollo. "I never really differentiated much between highways and freeways and interstates, I just called it a highway." The word "Ventura" captivated him. "I've seen that name in my mind's eye," he said. "Ventura. It was on the freeway signs. It was just trying to paint that picture . . . that vibe on the West Coast."

It was hard to escape "Ventura Highway" on FM radio in the 1970s. Even as an oblivious kid, I recognized Southern California in the song's groovy acoustic harmonies, though I was confused by the conflicting emotions it conjured up. A wistful loneliness, perhaps. A promise of freedom. Longing for something lost that maybe you didn't appreciate when you had it. As social media platforms reveal, the song still strikes a chord for listeners across generations, especially those who miss youthful drives along the freshly minted Ventura Freeway.

Its place secured in the cultural zeitgeist, the Ventura Freeway made the hit parade once again on March 10, 1985. Or, as the *L.A. Times* pointed out: "VENTURA FREEWAY—IT'S NOW NO. 1." But its position as a chart-topper was not an enviable one. That's because the Ventura Freeway had just passed the San Diego Freeway as the most heavily trafficked thoroughfare in Los Angeles. (The hit parade quip was by a Caltrans spokesperson, who marveled at its record traffic peaks.) After running neck and neck with the 405 every year, the 101 had finally edged its rival—267,000 average vehicles per day versus 266,000—based on the worst congestion spots for each freeway. For the 101, that spot was just west of the 405. For the 405, it was where it passes Santa Monica Boulevard. Interestingly, the Santa Monica Freeway—the busiest freeway in the world in the 1970s—dropped to third on the list of L.A.'s freeways, at 240,000 vehicles per day (at Crenshaw Avenue). For the last several decades, all three freeways have variously competed for the dubious honor of having the most traffic at their busiest points, though the 405 is currently the busiest overall.

As expected, Caltrans's electronic road sensors also ranked the 101's junction with the 405 as the most utilized interchange

OFF-RAMP ↗

The Adopt-A-Highway program—in which individuals or organizations lend time or money to maintain freeway shoulders in return for their name on a sign—started in Texas and was launched by Caltrans in 1989. For years, Bette Midler famously sponsored the 101 near Laurel Canyon Boulevard. The entertainer didn't personally do the cleanup (except while playing herself on *The Simpsons*). On the other hand, McDonalds employees donned work gloves and trash bags for an adoption near Newbury Park. Here's guessing they plucked out many discarded items from their own restaurant.

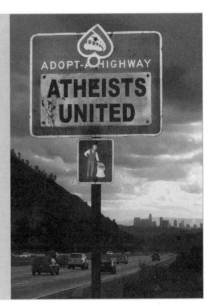

This "Atheists United" Adopt-A-Highway sign on the Glendale Freeway has been a fixture for years.

in the city, surpassing the East L.A. Interchange. Thus, the Ventura's traffic figures—both the freeway itself and its interchange with the San Diego Freeway—made it the busiest route in the world. Subscribing to the axiom that there's no such thing as bad publicity, Caltrans approached *The Guinness Book of World Records* in '85 with their results. David Boehm, an editor at *Guinness*, stressed that any claim for "No. 1 in the world title" would have to be backed up by an independent source based out of London. "We're particular about what we put in the book," he sniffed. Recent data suggests that the East L.A. Interchange is consistently the busiest. L.A. drivers, you be the judge!

The Ventura Freeway may have been snubbed by the record books, but it did receive one designation with surprising ease. In May 2017, the state Senate passed a resolution by Senator Anthony Portantino (D-La Cañada Flintridge) to name the four-mile portion between SR-2 and I-210 the President Barack H. Obama Highway. Obama's time at Occidental College from 1979 to 1981 inspired the senator, who related that the forty-fourth president commuted along the 134 from his apartment in Pasadena to the Eagle Rock campus. "He was a great statesman and president [with] a specific connection to this freeway," Portantino said. "California has a long history of designating our freeways as reminders of the accomplishments of important leaders."

The Obama Highway portion of the 134 starts at the freeway's junction with SR-2 and ends at I-210.

That was true, although the state often ended up with yolk on its face. In April 1971, the Marina Freeway (SR-90) was renamed the Richard M. Nixon Freeway during his first presidential term despite an unwritten policy that freeway honorees should be deceased. After Nixon resigned in disgrace three years later, the route was snatched away from him, reverting to the Marina Freeway. As of this writing, Obama's roadside signs are still up. If he can avoid scandal for the rest of his life, he will be the rare recipient who lives to die with his freeway intact.

And if not, no worries . . . we'll always have the long days of sunshine from "Ventura Highway," where the free wind is forever blowing through our hair.

Westbound traffic on the Ventura Freeway through Burbank's Media District at dusk.

Chapter 7

THE SAN DIEGO FREEWAY
Interstate 405 / Interstate 5
(1957–1969)

"Go big or go home."

It's a bit of a gaudy phrase, implying that nothing short of an all-out assault will do when sometimes subtlety is in order. But is there are more apt way to describe the San Diego Freeway?

The 405 reigns supreme as the busiest freeway in the world. It's the longest freeway in Southern California and was the costliest to build for its time. It has inspired a bonanza of nicknames, all of them epic-sounding. Construction through Sepulveda Pass was known as The Big Cut. Its junction with the 5 is called the Orange Crush. It's the only freeway to inspire cinematic labels. Carmageddon. Carmageddon II. Jamzilla. And don't forget the small screen. The 405 ruled there, too, costarring with O. J. Simpson and his white Ford Bronco during the live "slow-speed" chase of 1994—an event witnessed by almost 40 percent of the country. The freeway even dominates national social media. Who can forget the terrifying viral video posted by motorists driving through a mountainous tunnel of flames during the Skirball Fire of 2017? Because it invokes so many over-the-top qualities that we associate with Los Angeles, the 405 is *the* defining L.A. freeway—right down to residents' provincial belief that everything revolves around Los Angeles, including a freeway that isn't even named after it.

The San Diego Freeway's southern extremity is, of course, San Diego. From there, it cruises northward toward Orange County as I-5, switching over to the 405 at the "El Toro Y." The

405 rejoins the 5 in San Fernando to complete its 157-mile journey. The focus of this chapter will be the 48.5 miles of it that reside in Los Angeles County. As the most trafficked stretch of roadway in the Southland, it is also the most loathed. Billionaire commuter Elon Musk hated its perpetual gridlock so much, he thought he could bore his way beneath it. He got a mile in before giving up—or shall we say "giving in"? Because really, that's what 400,000 Angelenos do on a daily basis. The 405 is more than just a bad habit. It is the abusive lover you just can't quit. Every time you vow "never again," you come back for more. And like any dysfunctional relationship, you always hate yourself in the morning. Come to think of it, in the afternoon and evening, too.

Since L.A. drivers can't twelve-step their way out of it, it's high time to explore the powerful hold on Angelenos' psyches that is the San Diego Freeway.

OFF-RAMP ↗

Caltrans's official position is that the 405 ends in Downtown San Diego, where I-5 meets SR-94. The fifteen miles from Downtown to the Mexican border fall under the John J. Montgomery Freeway, which was completed in June of 1955. Google Maps recognizes this stretch as both the John J. Montgomery Freeway and the San Diego Freeway, and some people insist that the San Diego Freeway ends at Mexico. Choose your own reality.

Like that other fabled freeway pass through the Santa Monica Mountains—the Hollywood Freeway—Sepulveda Pass started as a foot trail for the Tongva people and was later used by the Spanish army. The pass was named after the Spanish soldier Francisco Xavier Sepulveda, whose family would claim several land grants. After California joined the union, the Sepulveda route was upgraded to a wagon road in the 1870s by Isaac Lankershim and Isaac Newton Van Nuys—two entrepreneurs from the Valley seeking a shortcut to the ports of Santa Monica.

Northerly view charting the San Diego Freeway from the coastal basin, through Sepulveda Pass, and into the San Fernando Valley.

But the road was not maintained and eventually deteriorated. Los Angeles at the turn of the century was largely consigned to the Downtown area, with Western Avenue essentially the western boundary and the San Fernando Valley not yet annexed. The pass's imposing incline—700 feet over 2.5 miles—also discouraged capital improvements.

Still, the city's westward migration necessitated reliable passages. In the 1920s, Sepulveda Boulevard became the main north-south axis linking San Fernando with San Pedro. Groundbreaking through the pass commenced in 1929, with Sepulveda Tunnel excavated shortly thereafter. (Almost a century later, this three-lane channel under Mulholland Drive remains exactly as it was then, despite fruitless attempts to widen it.) A predictable script followed: Street route proves popular. Route is declared inadequate. Route requires upgrading. How about a freeway?

A federally funded interstate through rugged Sepulveda Pass was an easy sell. Before its present-day alignment, Sepulveda Boulevard was a perilous road with hairpin curves that accounted for dozens of deaths—sixty-five during the 1950s alone. By then, 40,000 motorists were using it, many tending to aerospace or defense jobs in the South Bay or Valley. Northbound traffic regularly backed up to Sunset Boulevard.

The L.A. City Council recognized the need for a canyon thoroughfare as far back as 1943, when it endorsed the proposed Sepulveda Freeway that would parallel its namesake boulevard. After World War II, the freeway appeared on state planning maps as a "downtown bypass" to Interstate 5. Early on, the Highway Division purchased five million dollars' worth of property using "Chapter 20 Money," an internal term referring to advance acquisitions of vacant lands to forestall private construction. Up through the mid-'50s, the state spent an additional $12 million to buy out 15,000 home-dwellers—from the leafy enclaves of Sherman Oaks and Brentwood to the blue-collar belt of the South Bay to the agricultural tracts of Orange County. Like an exponentially

The San Diego Freeway at the start of 1962.

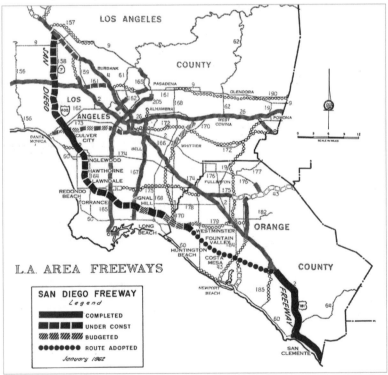

growing tapeworm, the free-way eventually mapped out at 93.7 miles, from San Fernando to San Clemente at the San Diego County line.

On September 21, 1954, one hundred onlookers watched Governor Goodwin Knight and members of the State Highway Commission don gloves and shovels to break ground on the freeway's first construction job: the Sunset Boulevard overpass. A few months later, the commission officially changed the name of the Sepulveda Freeway to the San Diego Freeway. To which the ghost of Samuel Morse could be heard telegraphing "What hath God Wrought?"

Early construction set the tone for future spectacles that would befall the freeway. As crews excavated the first 1.5 miles between Moraga Drive and Waterford Street, a trench-digger struck both gas and sewer lines, causing them to mix. "Gas fumes escaped and threatened explosion in hundreds of homes by backing up through plumbing fixtures," a local paper explained. Residents risked their lives if they lit a cigarette or had open pilot lights. As it was, the flammable sewer lines ignited nine homes and one business, setting off explosions around Sepulveda and Montana Avenue. Six square miles between Sunset and Washington Boulevards were cordoned off, and several people were rushed to the hospital with severe burns. One resident swore that his TV set exploded.

In another bizarre, high-profile case, two car thieves took a spin down the half-complete lanes. Police cars wailed behind the vehicle, leading to the first high-speed chase (note I didn't say "slow-speed") on the San Diego Freeway. The pursuit ended off of Wilshire Boulevard when the car slammed into a curb. One of the occupants tried to flee on foot, but the officers shot

A highway crew affixing a sign to the Sunset Boulevard overpass. California's original freeway signs were black with white lettering, with miles underlined in tenths. They were changed to green and white (with fractions for miles) to conform to 1962 federal standards that proved to be more reflective at night.

him, sending him to the hospital with flesh wounds. Turns out he was fourteen years old. When the cops asked him the name of his partner-in-crime, he told them to get lost, no stool pigeon was he.

The next day—March 30, 1957—coverage of the joyriding juvenile delinquent took a back seat to the opening of a two-mile segment of the freeway through Westwood. Three hundred people attended the dedication, including Governor Knight and Mayor Norris Poulson. In a spin on the golden spike, they used a golden saw to cut through a gold-painted wood barricade.

THE BIG CUT

Following the highway division's policy of prioritizing traffic needs and right-of-way clearances, a checkerboard of sections opened over the next few years: 1.2 miles from Valley Vista to Burbank Boulevards; 2.3 miles between Venice and Jefferson Boulevards; and 7.7 miles from Capistrano Beach to San Clemente, the original southern endpoint. But these were mere opening acts for the much-anticipated main event . . .

Grading work for the Big Cut through Sepulveda Pass in 1961. The cut was so deep, one newspaper boasted that a 30-story building could be hidden in the man-made gorge. Sepulveda Boulevard, on the right, would relocate west of the freeway upon its opening.

The Big Cut through Sepulveda Pass was the state's most ambitious freeway excavation to date. With a record contract of $17.2 million, the Public Works Department deemed it the largest earthmoving project in California history. The scope of it all was daunting: the removal of 18 million cubic yards of earth, 90,000 cubic yards of concrete, and 6 million pounds of steel. The Big Cut itself was nothing less than gouging out a brand-new canyon from scratch that would measure 1,800 feet wide and 300 feet deep at its summit. The to-do list also included building Mulholland Bridge and realigning Sepulveda Boulevard. Under pressure to work quickly, the state announced it would complete the 5.7-mile, eight-lane strip from Moraga Drive to Valley Vista Boulevard in two years.

Tractors started carving up the pass in August 1960. Engineering challenges abounded, to say nothing of the rattlesnakes. The steep slopes of the canyon were notoriously unstable, requiring massive retaining walls and a complex system of drainage culverts. To make the freeway viable for trucks, the Valley side was sliced down from 1,400 to 1,175 feet, creating a 5.5 percent grade for a mile and a half. South of Mulholland Drive, the grade was a more gradual 3 percent. Lowering the mountaintops produced a strange phenomenon. Residents at the north end of the pass swore that ocean breezes were now "ventilating" their neighborhood. Max Bankert, the U.S. Weather Bureau chief in Los Angeles, wasn't buying it. "It's like getting water to run up hill," he explained. "The cool air layer on the south side of the Santa Monica Mountains is heavy and very shallow. It rarely gets thick enough to filter through that canyon." Dr. Robert B. Lamb, a professor of geography, was more blunt: "People just don't have any conception of the magnitude of weather." But if the wishful thinking of cooler weather made folks in Sherman Oaks happy, who were scientists to argue?

Sepulveda Pass wasn't the only section met with urgency. In September 1960, civic leaders pushed the California Highway Commission to hurry up the freeway near Los Angeles International Airport. During the 1950s, Los Angeles's population soared almost 50 percent, to 2.5 million. The former Mines Field grew along with it, boasting Pereira and Luckman's space-agey Theme Building. Current roadways were ill-equipped to handle the crush of travelers, much to the despair of Francis Fox, the airport's general manager. "We will have the finest airport in the world, but it may not be used because of poor surface transportation," he warned. "The one thing without which we cannot operate is the San Diego Freeway." Fox got his wish in the state's 1961-1962 highway budget—$25 million to pave ten miles of

The freeway as payday: That was the pitch by realtors in Carson and other communities, who advertised the impending San Diego Freeway as a selling point.

freeway from Jefferson Boulevard to Torrance. Work began in January 1961 and was
completed that year.

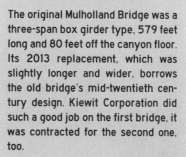

In the winter of 1961-
1962, above-average rainfall
triggered mudslides in northern Sepulveda Pass. Nine cliffside homes on Royal Ridge
Road and Scadlock Lane were
seized so contractors could
shave down and reinforce the embankments. The inconvenience
added another $2.1 million to the budget (mostly paid by the feds,
who were footing 91.6 percent of the freeway's bill) and delayed
the pass's opening. As summer slipped into fall, the press broadcast almost-daily dispatches from the site. After all the lanes
were poured with concrete—leaving only the construction of
shoulders, curbs, and gutters—one engineer said, "Just putting
the lace on the dress."

The freeway was fully dressed by December 21, 1962. The
Public Works folks released a newsreel called *A Special Day*,
which covered the opening day festivities. It's a classic elementary school-type documentary, with warbly patriotic music and a
stentorian narrator as pee-wees in strollers and smartly dressed
parents watch a procession of square highway men take turns at
the lectern. One of them draws tentative applause after he ex-
horts, "Take care of each other on this highway . . . let's have no
accidents and nobody killed on this thing." Governor Pat Brown
flew down for the dedication, but it is Los Angeles Mayor Sam
Yorty—ranked by political scientists as the third-worst mayor of
a U.S. city from 1960 to 1997—who grabs the spotlight. Near the
Sunset overpass, he hops into an orange Bell helicopter on loan
from the fire department. Its landing gear snaps a ceremonial red
ribbon held by a Jackie O lookalike as it lifts into the air. Then, as

Los Angeles Mayor Sam Yorty and a Nisei Week Queen in 1967. Yorty believed freeways were the building blocks that would make L.A. a world-class city. But the populist leader courted controversy with outlandish stunts and comments that evinced racial baiting.

it buzzes over the Mulholland summit, Yorty partakes in another ribbon-cutting, this time extending his arm out of the cockpit and scissoring the ribbon from the air.

At 11:20 a.m., police unshackled the southbound lanes. Within minutes, the first speeding ticket was issued to a Ronald P. Tamkin from West L.A. The northbound lanes were not due to open until 3:00 p.m., but after antsy motorists threatened to burst through the barricades, officials removed them early. As cops chased down speed demons, others suffered flat tires, overheated engines, or traffic accidents (seventy-five more wrecks would occur over the weekend). The freeway free-for-all resulted in an immediate SigAlert. Unlike the future

OFF-RAMP ↗

In March 1963, Councilman Karl Rundberg accused Mayor Yorty of "pulling one of the phoniest publicity stunts by using a Fire Department helicopter to lift him to a ribbon-cutting ceremony . . . for the opening of the San Diego Freeway." At a finance committee meeting, he asked Police Chief William H. Parker if Yorty abused his power by using the helicopter as his own "taxi." Though a freeway pageant could hardly constitute an emergency, Parker demurred. "I don't believe it is within my power to challenge whether he is on emergency business," he said.

The freeway transitioning from the Sepulveda Pass to West Los Angeles, facing south.

bust that was Carmageddon, this one truly lived up to that moniker. Or, in keeping with mid-century cinema, maybe it was more of a War of the Wheels.

Once things settled down, however, a freeway through Sepulveda Pass turned out to be a godsend. Even with 100,000 vehicles a day—double the volume that Sepulveda Boulevard saw—commuters could now save twenty to twenty-five minutes traveling between the Westside and the Valley. As the *Los Angeles Times* noted, the juxtaposition of Sepulveda alongside the San Diego Freeway created "an ironic before-and-after ad, one illustrating man's engineering progress and the other hinting at the shortness of his vision."

Two decades into L.A.'s freeway-building, collective data reinforced the mode's relative safety compared to surface streets. The Auto Club of Southern California calculated that the metro area's 252 miles of freeway would save 250 lives in 1962, while the state estimated that the embryonic 405 had spared nine people from untimely deaths on Sepulveda. Motorists would save money, too—$81 million in operating costs, $60 million in collision costs, and, thanks to newfound efficiency, $224 million in time saved. This last figure is downright laughable by today's 405 standards. "I wish I could be here about five o'clock this afternoon," the unknown narrator marvels at the end of *A Special Day*. "I'd like to see the commuters coming home from their jobs on the other side of the mountain!" Yeah, no.

The San Diego Freeway's planned route through the Valley, facing south, 1956.

Sepulveda Boulevard, meanwhile, underwent an image makeover. As vehicles diverted to the San Diego Freeway, the previously maligned frontage road suddenly fell back into favor. Newspaper columnist Matt Weinstock found it "unbelievable that in such a short time a highway that used to throb day and night with traffic can be so deserted, so isolated." A subdivision called Royal View Estates played up the street's proximity to its homes: "Presto! Sepulveda Boulevard as it was 30 years ago" read one ad. "The new San Diego Freeway siphons off all the heavy traffic!" It was a clever inversion—touting an old street over a brand-new freeway. Clearly, if L.A.'s freeway system wasn't quite at its saturation point yet, it was getting pretty darn close.

REBELS WITH A CAUSE

Like other freeways built during the changing times of the 1960s (or later), the burgeoning 405 Freeway served as a Rorschach test for the communities it traversed. The freeway was a key catalyst for the suburbanization of the Valley, which grew twice as fast as the rest of L.A. from 1961 to 1966. Once it reached the city of San Fernando (its northern terminus at I-5), a trip that used to take ninety minutes from Long Beach was reduced to just under an hour. In denser, airport-adjacent neighborhoods like Inglewood—where 427 homes had already been demolished—residents were more wary of the freeway. Besides noise and air pollution, parents expressed concerns about children walking to schools near the raging thoroughfare. At the dedication of the La Tijera Boulevard exit, picketing mothers and their kids held up placards beseeching the state to build a pedestrian bridge over the 405 so students could avoid crossing the on- and off-ramps. The protests lasted seven months. Despite his proven record as a safety advocate, County Supervisor Kenneth Hahn was unable to secure additional funding for a second overpass, though he did arrange for protective fences to be installed along the sidewalk curbs leading to La Tijera's access ramps (the rusty relics remain to this day).

Not every community's protests were so virtuous. In Westwood, thousands of UCLA students—many of them drunken frat boys—stormed the freeway near Wilshire Boulevard, blocking traffic, screaming obscenities, and starting bonfires. Their beef?

Parents and their children protesting the freeway's incursion through their Westchester neighborhood in 1963. Their pleas for a separate pedestrian crossing over the freeway went unfulfilled.

Late in 1966, UCLA students took to the 405 to demonstrate against rival USC getting into the Rose Bowl... or at least, that's what started it. By this point, a few appear to be saying, "Dude, we're walking on the *freeway!*"

The NCAA's surprise decision to select the USC Trojans to play in the 1967 Rose Bowl instead of the UCLA Bruins. Police invoked an unlawful assembly and arrested thirty people. By midnight, the rioters dispersed after most of them probably forgot what they were even protesting against.

As a stark contrast, consider another protest farther down the freeway at UC Irvine. On December 6, 1968, the university's Black student body called for an end to on-campus verbal harassment, the hiring of more Black professors, and investigations into white landlords in neighboring Orange County cities who refused to rent to African Americans. The same day they aired their grievances—a mere one mile away—officials dedicated the southern end of the 405 at Jamboree Road with a celebration steeped in gauzy nostalgia. It featured an 1885 Wells Fargo

OFF-RAMP ↗

When the northern end of Interstate 405 opened, it was blessed by a priest from San Fernando Mission, which lies half a mile east of the freeway. Just north of the mission, a different faith wasn't as receptive. Representatives of Eden Memorial Park—where Groucho Marx and other Jewish luminaries are buried—secured a temporary injunction to block contractors from paving over an eleven-acre section of the graveyard.

stagecoach rumbling down the Jamboree on-ramp. On board was a gaggle of Orange County beauty queens—Misses Orange, Tustin, Fountain Valley, Newport Beach, Orange County, Santa Ana, Garden Grove, and Westminster. The coach was "held up" by two horsemen play-acting notorious robbers from 1856 who fled into a nearby canyon, freeing the damsels to witness the ribbon-cutting. Top *that*, L.A. Mayor Yorty.

The next day's *L.A. Times* banner: "FINAL 8-MILE LINK DEDICATED: SAN DIEGO FREEWAY NOW REALITY." The headline was a bit misleading. The eight-mile patch from Jamboree Road to Interstate 5 wouldn't open until January 1969—a month later. Also, while the San Diego Freeway did terminate at the Orange/San Diego County line, it would eventually extend all the way to San Diego. Nonetheless, the temporary "final" numbers were impressive enough: 93.7 miles built in thirteen years at a cost of $355 million (150 miles at $555 million, once one added

I-5's completion in San Diego later in '69). Just as it had for the northern Valley, the 405 helped drive development in southern Orange County. "Could the South Coast become San Fernando-by-the Sea?" one newspaper asked. In fact, the freeway wasn't even finished yet when the state commenced the first of many widening projects at the dreaded "El Toro Y"—where the 405 meets the 5—a perpetual choke point dubbed by traffic reporters as "the Orange Crush." But now was not the time for criticism. As

OFF-RAMP ↗

Older Southern Californians may remember the 405-adjacent Lion Country Safari, where families could drive through 140 acres of Irvine parkland and encounter wild animals (to keep people in their cars, signs said "No Trespassing: Violators Will Be Eaten"). Imagine a poorly run zoo without cages. A kid was mauled by a Siberian tiger, a hippo named Bubbles terrorized Laguna Beach after it escaped—three different times, once for nineteen days—and an elephant named Misty killed a game warden before stomping off toward the San Diego Freeway. Police shut down the freeway for three hours and evacuated businesses. To Irvine's great relief, the park closed in 1984.

the '60s wound down, the *L.A. Times* was drunk on the possibilities pitched by the freeway lobby, namely a limitless paradise of twenty-nine other freeways honeycombing the Southland over the next decade, so many that the "landscape will look from the air like an old woman with varicose veins."

Unlike that unfortunate turn of phrase, the speculative freeways were eventually deleted.

NEXT STOP: CRAZYTOWN

Two years after its completion, the northern end of the San Diego Freeway narrowly escaped a catastrophe for the ages. As previously mentioned, on the morning of February 9, 1971, the 6.6-magnitude Sylmar Earthquake led to death and destruction on a couple of interstates. Wire services initially placed the killing of two motorists from a collapsed bridge at the San Diego-Golden State Interchange. This tragedy actually occurred at the Foothill-Golden State Interchange, though the 405 did suffer some bridge damage where it meets the 5.

The junction of these two freeways was directly southeast of the Lower Van Norman Reservoir—L.A.'s largest at the time.

On the brink of a real-life 1970s disaster movie: The Lower Van Norman Reservoir, one day after the 1971 Sylmar Earthquake. Half of the dam's concrete facing has slid into the water, which sits alarmingly high against the earthen portion—this *after* 24 hours of frantic drainage. Hundreds of thousands of Valley residents lived below this dam.

Its 1,100-foot-wide earthen dam overlooked north Valley communities like Granada Hills and Northridge. As the quake rumbled through, the top of the dam began to crumble, leaving 3.6 billion gallons of sloshing water only six feet below the lip. Fearful of aftershocks, authorities ordered the immediate evacuation of 80,000 people in a twenty-square-mile swath south of the dam and west of the 405.

OFF-RAMP ↗

In 1968, Peter Bogdanovich made his directorial debut with *Targets*. The movie's most chilling sequence involves an alienated young man casually eating a sandwich atop an oil storage tank off the 405 in the Valley. He then pulls out a rifle and randomly fires at passing cars on the freeway, causing several to skid and crash. The scene was inspired by a 1965 incident in Santa Barbara County, when a sixteen-year-old hilltop sniper killed three people driving on the 101 Freeway.

According to *California Geology* magazine, if the shaking had gone on for just two more seconds, the dam would have given way. The resulting flood would have submerged the San Diego and Simi Valley Freeways and, according to a UCLA study, killed between 71,000 and 123,400 people—by far the worst natural disaster in American history. (The reservoir is now an empty flood control basin.)

By the 1980s, the San Diego Freeway's reputation as a living nightmare was pretty much cemented. Statistics bore out drivers' frustrations. It hosted two of the most congested junctions in the United States—at the 10 and 101 Freeways. To survive the San Diego Freeway, it was best to laugh to keep from crying. "The freeway was numbered '405' because traffic moves at four or five miles per hour" went one joke. Its traffic became a punchline on late-night comedy bits. On May 10, 1994, David Letterman connoted the freeway with "crazy" in one of his Top Ten Lists—a sure sign that you've arrived on the national scene:

TOP TEN SIGNS THAT RICHARD SIMMONS IS NUTS
10. Recently found naked on San Diego Freeway playing solitaire with his Deal-A-Meal cards.

Coincidentally, one month after the farcical Richard Simmons meltdown, the 405 served up a real-life case of a celebrity gone mad—the "slow-speed chase" involving O. J. Simpson. Witnessed by 95 million Americans, it was the first "Where Were You?" moment to be played out on a Los Angeles freeway.

On the morning of June 17, 1994, Orenthal James Simpson was ordered to surrender to Los Angeles prosecutors on charges of murdering ex-wife Nicole Brown Simpson and her friend Ron Goldman. But after giving his lawyers the slip, Simpson jumped into a white Ford Bronco with his buddy Al Cowlings behind the wheel. That afternoon, the Bronco materialized on the Santa Ana Freeway, where authorities spoke with O. J. over his cell phone. The legendary football star and actor claimed to be suicidal. He held a gun to his head and said he only wanted to see his mother. At least a dozen highway patrolmen gave the Bronco a wide berth, clearing the freeway to safely escort the SUV as it cruised along at around forty-five miles per hour.

It was the type of unfolding saga that was custom-made for Southern California. Live freeway chases from the air had become the norm on L.A. newscasts, and stations were ready to kick into action. By the time Cowlings's Bronco had turned onto the San Diego Freeway, virtually all the local stations' helicopters were zeroed in. Nationally, NBC Sports memorably cut to the chase in a split screen with the NBA Finals, and national broadcasters including CNN aired it live through local feeds. "There's no rule book on this because this has never happened," said Larry King, who interrupted his own show to call "play-by-play."

KTLA Channel 5 News had a seasoned team led by Hal Fishman, an accomplished pilot who knew L.A.'s freeways in ways that other anchors didn't. He was among the first to surmise that O. J. was traveling to his estate on Rockingham Avenue in Brentwood. His banter with KTLA's Skycam 5 Pilot, Craig Dyer, during the fugitive's home stretch made for riveting television.

DYER: "Currently approaching the 405/10 Interchange, north-bound..."

FISHMAN: "From the I-10 Interchange up to Sunset Boulevard is only going to be maybe three or four minutes more."

DYER: "Oh, most, definitely. The off-ramps are quite blocked."

By this point, Angelenos who had caught the live chase at home were clogging the 405's bridges and freeway shoulders, rooting O. J. on—waving hands, blowing kisses, honking horns.

DYER: "A good number of people are pulled on the right shoulder, and are out of their vehicles, watching this procession proceed up the 405 freeway here... watching this as it passes them by."

FISHMAN: "You should be approaching, pretty close to Wilshire Boulevard, aren't you?"

DYER: "Coming up on Santa Monica Boulevard at this time."

FISHMAN: "There are only three off-ramps now remaining if he's going to that Brentwood estate. There's the one coming up on Wilshire, which you're probably approaching now."

On the morning of June 18, 1994, newspapers like this one in New Jersey and across America splashed officers' freeway pursuit of Simpson's Bronco on their front pages.

DYER: "That's affirmative, Hal."

FISHMAN: "Okay, and then right after that, probably another thirty to forty seconds, you'll be at Waterford. We'll see if he turns off on Waterford, but he's in the left lane!"

He meant the Montana Avenue exit; Waterford Street (now closed) was only accessible from the southbound lanes, but we'll cut Hal some slack here due to all the chaos.

FISHMAN: "Okay, now it's crucial to see if he starts moving over to the right lane."

DYER: "It doesn't appear that that will be the case due to the number of vehicles blocking those right lanes!"

As the Sunset Boulevard exit loomed, the Bronco was still in the fast lane. This was a big moment of suspense: Would it leave the freeway at Sunset, which would indicate O. J. was headed home? Or would it keep going north?

At this point, KTLA cut to audio from a field reporter, Eric Spillman, who had managed to get into a car just behind the Bronco.

SPILLMAN: "The Sunset Boulevard off-ramp is open, and it appears, uh, that they think he might get off here."

He did. As the vehicle turned left onto Sunset, hordes of O. J. fans rushed across the Sunset overpass—the first one built over the freeway.

SPILLMAN: "It was an amazing situation on that overpass there . . . I don't know how he's going to be able to drive through all those people."

For the next several miles, Cowlings lived out a Super Mario Kart video game, swerving through traffic and red lights while civilian motorcyclists zipped behind him. When the Bronco finally pulled into the Rockingham driveway—culminating sixty miles along freeways and city streets—dozens of supporters showed up at the edge of Simpson's property, bearing signs such as "WE LOVE THE JUICE" and "THE JUICE IS LOOSE." As night fell on Brentwood, O. J. remained in the backseat, communicating with authorities over the phone while, outside his window, he was serenaded with chants of "Juice! Juice! Juice!" The Juice was taken into custody at 8:47 p.m.

Given the legal future that awaited O. J. Simpson—a botched guilty verdict in the criminal trial, a $33.5 million liable judgment in a civil trial, a nine-year prison term for armed robbery and kidnapping—it was not Los Angeles's proudest moment. But, for better or for worse, it *was* a quintessential L.A. moment—one only made possible thanks to its signature freeways.

BLOATED CONSTRUCTION BLOCKBUSTERS

Meanwhile, the most enduring feature of the San Diego Freeway—mind-numbing traffic—continued to worsen. Built to

OFF-RAMP ↗

Perhaps the most controversial signs ever placed on a freeway were on the San Diego Freeway—the stretch that goes through Camp Pendleton and the Border Patrol Checkpoint. After forty undocumented immigrants were struck and killed crossing the 5 in 1989 and 1990, Caltrans graphic artist John Hood came up with a striking visual to warn drivers: giant yellow "CAUTION" signs with the silhouettes of a sprinting father, mother, and little girl, pigtails flying. Despite good intentions, the signs were eventually removed for their dehumanizing aspect, essentially equating people with livestock.

One of many "Immigrants crossing" signs since removed in northern San Diego County.

handle 200,000 daily vehicles, the San Diego-Ventura Freeway Interchange was squeezing out over 500,000 in 2001. A replacement would require condemning dozens of homes and businesses at a cost of $1 billion. Since that was not feasible, Caltrans elected to spend $50 million on improvements, which even a former commissioner admitted was merely "putting a Band-Aid on a bleeding artery." Still, the main problem was the volume of vehicles on the entire freeway. Caltrans's internal rating system tagged the 405 with not just an "F" but an "F2"—essentially, a double-failing grade. Even the newly opened Getty Center above Sunset owned up to the problem in their marketing materials: "Coming here is as easy as heading up the 405. But don't let that discourage you."

The simplest way to alleviate traffic was to install carpool lanes. This fit into MTA's grand plan of doubling the number of HOV lanes by 2015. Notwithstanding an abortive attempt in L.A. County that was halted by Governor Jerry Brown in 1976, Orange County implemented the first 405 rideshare lanes in the late 1980s. But an early stretch in Costa Mesa wasn't universally embraced. Its opening was delayed when a group called the Drivers for Highway Safety threatened to sue Caltrans for not thoroughly assessing the efficiencies of high-occupancy lanes. At one point, a donnybrook ensued between carpool backers and detractors on an overpass in Irvine. A CHP officer broke it up after the warring parties disrupted traffic.

Perhaps owing to the county's strong bent toward individualism and personal privacy, many motorists resented the use of video cameras to catch carpool violators. Besides the creepy Big Brother aspect, what if the cameras caught a mistress in a male driver's passenger seat? Or failed to capture a second passenger who was out of view, perhaps asleep in the back? In 1996, SR-73—the first publicly operated toll freeway in the Southland—finally opened as O.C. residents grew inured to video-monitored roadways.

In the early 2000s, MTA provided 90 percent funding for a 7.8-mile, $20 million southbound carpool lane through Sepulveda Pass. Engineers were able to realign the freeway's existing median to execute the job. Adding a northbound lane, however, would require physical widening of the canyon walls. It would also necessitate replacing the Mulholland and Skirball Center Drive overcrossings. This presented obvious challenges. Initial work would require shutting down the freeway for fifty-three hours. Officials planned it around a summer weekend in July 2011. Of course, depriving L.A. drivers of their birthright was a risky proposition that required getting the word out well in advance. Social media offered the perfect platform for pithy updates using hashtags and humor.

It was Los Angeles Councilman Zev Yaroslavsky who coined the term "Carmageddon," a clever portmanteau of "car" and "Armageddon." The impending closure now had a name. Not just any name, but a relatable one that conjured a cataclysm to sufficiently scare people away from the freeway while also winking at our love of disaster movies—a menu that included *Armageddon*. The countdown to construction ("premiering" on a Friday,

Even Sigalert.com caught Carmageddon fever. The normally staid traffic website used the word in conjunction with the 405's closure.

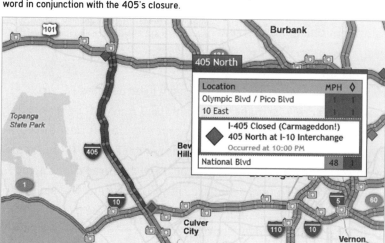

natch) was reminiscent of "Coming Soon" trailers. I remember having no reason to use the 405 Freeway that weekend, but even I got caught up in the mania . . . and the dread of other freeways across L.A. becoming overloaded from the closure.

And then, a weird thing happened. Motorists stayed off the freeways, and the work finished seventeen hours early. Carmageddon was a bust. No, wait, it was a hit! That is, it all went better than expected. Mayor Antonio Villaraigosa praised L.A.'s citizens, and Caltrans came out smelling like a rose—so much so, it did what any Hollywood studio does when it experiences success. It announced a sequel.

Carmageddon II was billed for the last weekend of September 2012. The main objective was to finish tearing down the old Mulholland Drive Bridge (only its southern lanes were demolished in 2011, allowing the northern lanes to become two-way while the replacement was being built). Villaraigosa guarded against public apathy. "Last year, we proved that Angelenos far and wide could

Mayor Antonio Villaraigosa and Councilman Zev Yaroslavsky in 2011. It was the councilman who coined the term "Carmageddon." Officials estimated that LA Metro received the equivalent of $50 million in advertising from all the media coverage.

rise to the occasion and cooperate with authorities to turn Carmageddon into 'Carmaheaven,'" said the mayor (cue eye rolls). "It was truly one of L.A.'s finest moments. Do not become complacent." Caltrans placed a "Countdown to Closure II" billboard off the 405. News stations assaulted us with more groaning movie puns: "This is one case where they do not want the sequel to live up to its hype!" "Will Carmageddon Sequel Be Worse?" Will it be "The Traffic Strikes Back"?

Once again, the work finished early and everyone somehow survived the weekend. Construction of the carpool lanes continued without major closures until Presidents' Day weekend in February 2014. That's when MTA trotted out "Jamzilla." "We wanted to come up with a term that would be like Carmageddon in its ability

OFF-RAMP ↗

The opportunity to experience an empty freeway proved too good to pass up for Matt Corrigan. During Carmageddon, the thirty-year-old Sherman Oaks resident gathered two friends for a formal candlelight dinner in the middle of the northbound lanes. The resulting photo went viral, though some questioned why news outlets were "rewarding" trespassers with free publicity.

to influence the public," explained a spokesman for Metro. "If you loved Carmageddon, then get ready for Jamzilla!" the *L.A. Times* announced. Maybe it was the overkill of movie jargon, or the attempts at yet another derivative spin-off, but public cynicism was sky-high leading up to the (hopefully) final installment of the trilogy. As it turned out, the freeway was kept partially open as contractors finalized the rideshare lanes. Completed

Crews removed 9 million pounds of concrete from the Mulholland Bridge demolition.

From *Carmageddon* to *Backdraft*, the San Diego Freeway has had numerous brush fires break out along Sepulveda Pass. Motorists willed themselves through walls of flames during the Skirball Fire of 2017, sharing their horrifying drives on social media.

in 2015, the widening of the 405—with HOV lanes now in place from the Santa Monica Freeway to the Ventura Freeway—had reached a staggering $1.6 billion. The project also led to new retaining walls, which some critics derided for their cheesy ersatz-rock patina, especially when placed alongside plain concrete sound walls.

So what did Angelenos get for their share of taxpayer money, which equaled the GDP of Belize? Worse traffic. That was the finding from INRIX, a traffic data firm. Despite two additional lanes, congestion increased during rush hour. In 2015, the average speed was twenty-eight miles per hour. In May of 2019, it dropped to nineteen miles per hour. Driving time through the pass went from twenty-three minutes to thirty-four minutes—a 48 percent bump. As someone who commuted on the 405 during those years, I can attest to these numbers. Even Zev Yaroslavsky expressed buyer's remorse. "I doubt the project would have been undertaken in the first place if we'd known it would cost $1.6 billion," the councilman said of his district's pet project. "There's a lot of bad taste in my mouth about this . . . It was a nightmare of a project." Once again, more lanes proved the fundamental theory of "latent demand," whereby any short-lived benefits are offset

Over one hundred million vehicles drive—or crawl—through Sepulveda Pass every year.

by those newly encouraged to drive. Also, one cannot overlook the advent of ridesharing apps like Uber and Lyft and an improving economy, both of which contributed to increasing traffic across the Southland starting in the mid-2010s.

Like a movie that eventually finds its audience, Carmageddon lived up to its hype after all.

SAVING THE 405

The San Diego Freeway consistently ranks as the busiest highway in the world—a claim that will likely extend well into the future. But there are some transit projects that could lessen the load. Thanks to Measure M—a local sales tax increase passed in 2016—Metro has committed to fast-tracking a rail line from the Valley to the Westside. (Once again, a monorail line—that sexy mode of transportation that always resurfaces—has also been considered.) A train along the 405 was proposed as far back as 1976, when backers envisioned the Sunset Coast Line from San Fernando to the Long Beach Freeway. The cost would have been $622.1 million. As of this writing, $5.7 billion is slated for a shorter route through Sepulveda Pass. Metro has allocated another $3.8 billion to extend rail service to LAX by the 2050s.

The agency is also exploring an ExpressLane—a fancy word for toll carpool lanes, which allow solo drivers to pay a sliding fee based on congestion conditions. Like it or not, these are the wave of the future. Measure M has earmarked several hundred million dollars for ExpressLanes on the 405 between the 101 and 10. It will be interesting to see how these new HOT (High-Occupancy Toll) lanes meet the standards set by the Federal Highway Administration. In order to remain in good standing for federal funds, interstate HOV/HOT lanes must maintain certain speed thresholds, otherwise they are considered "degraded" and at risk of losing aid.

No matter what pans out, the San Diego Freeway will remain the dragon that motorists must slay daily. It holds the dubious distinction of hosting the busiest interchanges in San Diego County (I-5/I-805), Orange County (I-405/I-5, aka the Orange Crush, aka the El Toro Y), and two of the busiest in L.A. County (I-405/I-10 and I-405/U.S. 101). A stand-up comedian once argued that you haven't really busted your Los Angeles cherry unless you've sat in traffic on the 405. At the risk of sounding delusional, I'll offer a contrarian take: there's something almost comforting about knowing the 405 will always be busy. It means our economy is humming. Life is pulsating. People are going places. (No pandemics!) Every once in a while, when I'm mired in traffic, I'll find random moments of beauty: The streak of white headlights and red taillights snaking through Sepulveda

Pass at night. The Valley unfolding before your eyes as you crest Sepulveda Pass from the south during the daytime. The rush of the ramps between the San Diego and Santa Monica Freeways, an interchange that architectural historian Reyner Banham described as a "kinetic experience as one sweeps through it." On the flip side, is there a more serene patch of freeway than the stretch where the San Diego Freeway brushes the Pacific Ocean, between San Onofre Beach and Oceanside? It's the rare interstate where cavorting dolphins just may accompany your drive.

But these are small diamonds in the rough. At the end of the day, there's no escaping the cold hard truth: the San Diego Freeway really does suck.

Proof that the 405 Freeway is toxic: Workers don hazmat suits to clean up molten sulfur that spilled out of an overturned tanker truck near Culver City.

Chapter 8

THE GLENDALE FREEWAY
State Route 2
(1958–1978)

Is there such a thing as freeway utopia? If there is, the Glendale Freeway just might be it.

The 9.3-mile strip from Silver Lake to La Cañada Flintridge consistently scores high marks from Angelenos. It possesses all the ingredients you want in a superhighway: light traffic, even during rush hour; clean lines and few curves, allowing for safe high speeds; majestic views in both directions (going southbound, the Downtown L.A. skyline; northbound, a trio of mountain ranges); virtually no graffiti; and most importantly, functionality—expedited access to communities northeast of Downtown.

Of course, one person's paradise is another's hell—or at least purgatory. Initial plans called for the Glendale Freeway to intersect with the Hollywood Freeway. Caltrans bought up hundreds of properties between Vermont Avenue and Glendale Boulevard to prepare for it. Suffice to say, transportation officials are not ideal landlords—it's their way *and* the highway. Even when the 2 Freeway's extension was erased from the books, tenants spent years trying to buy back their homes while the agency dragged its feet. Only an innovative initiative in 1979 helped return the domiciles to their original owners, providing a legislative guide for other lower-income neighborhoods across the city.

With one problem solved, another remains. Caltrans always intended for the terminus at Glendale Boulevard and Allesandro Street to be temporary, but it never designed a permanent solution. As a result, southbound freeway traffic soars onto surface streets like a flume screaming down a log-ride chute. Bless those

The Glendale Freeway. about five years after its completion. heading toward Downtown L.A.

who dare to walk or drive across Glendale Boulevard at the base of "Silver Lake Falls." It's either the most treacherous or thrilling end-of-freeway spot in the entire L.A. freeway network, depending on your point of view.

No such hair-raising moments exist for the rest of the route, although for much of the 1970s, it seemed that the actual *Glendale* portion of the Glendale Freeway would never get finished. When it finally was, L.A.'s love affair with the 2 officially began—so much so, bloggers have serenaded this mass-transit mistress with open love letters (while penning Dear Johns to her congested cousins).

But every relationship comes with conditions, and before plunging into one with the 2, first we'll have to get past its considerable baggage.

It's a timeworn cliché that just happens to be true: Southern California is one of the few places in the world where one can surf in the morning and snow ski by lunchtime. But pulling off this nifty trick means spending most of your time slogging up a two-lane highway through Angeles National Forest.

Now imagine if there were a freeway that took you straight from Santa Monica to, say, the slopes of Wrightwood. Highway planners imagined it, too. Once upon a time, a freeway would have started on Santa Monica Boulevard just west of the San

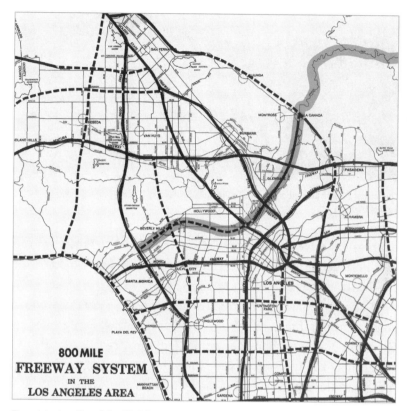

The original routing of the SR-2 Freeway had it starting at the 405 and ending in Angeles National Forest under three different freeway names. Only the Glendale Freeway portion was built.

Diego Freeway and crossed over to Silver Lake. From there, it would travel north to Glendale, slice through the San Gabriel Mountains, and end at Highway 138 in San Bernardino County. I know, it's a bit convoluted. But it all makes sense when you consider that planners often ran freeways over or alongside heavily trod highways. In this case, the proposed ninety-mile route would have mirrored the path of State Route 2. Consequently, an SR-2 Freeway would have followed suit under three different names that changed based on geography—the Beverly Hills Freeway, the Glendale Freeway, and the Angeles Crest Freeway. To the chagrin of SoCal boarders, dreams of a sand-to-snow

OFF-RAMP ↗

Glendale Boulevard has long been an important artery. The first subway in Los Angeles ran from Downtown to the boulevard in 1925. The street remained a major thoroughfare for vehicles and Red Car trolleys for decades, creating an early need for the Glendale Freeway.

The former Pacific Electric tracks along Glendale Boulevard, leading to the subway's Belmont Tunnel.

expressway never materialized. Only the Glendale Freeway—the part between Silver Lake and La Cañada Flintridge—made it off the drafting board.

Prior to 1955, the eventual freeway along SR-2 appeared on some planning maps as the Allesandro Freeway, a reference to Glendale Boulevard's former name (and the street that currently parallels the freeway to Riverside Drive). On September 1 of that year, the California Highway Commission officially designated it the Glendale Freeway "to forestall confusion concerning the freeway routing." Curiously, only four miles of the freeway traverse Glendale, and its destination is actually La Cañada Flintridge. But as the fourth most populous city in Los Angeles County, Glendale has earned its rightful name on Caltrans's iconic green-and-white signs.

As with the Foothill Freeway and others, officials chose to break ground at a troublesome

In the early days of freeway-building, some locals looked to make a buck off thirsty or hungry crews. These boys offered lemonade for 5 cents a cup in 1960, though other residents were known to gift them pastries and other goodies.

The sleek Glendale-Golden State Freeway Interchange, completed in 1962, facing southwest. Silver Lake Reservoir is visible in the upper right.

bottleneck—in this case, a 1.5-mile stretch between the Los Angeles River and Eagle Rock Boulevard on Fletcher Drive. Southern Pacific trains crossed at street-level here, producing, per a traffic engineer, "a gruesome track record." The Division of Highways' solution was to construct an above-grade freeway over the railway. In the process, it proposed turning Fletcher into a dead-end street on both sides of the tracks. City officials balked. Ditto Van de Kamp's Bakery, which employed hundreds of people and perfumed the neighborhood with the seductive aroma of fresh cinnamon bread. The idea of decommissioning a busy east-west corridor like Fletcher exposed the state's poor grasp of local traffic habits and its flawed belief that freeways were a quick-fix panacea. The state caved, agreeing to dip Fletcher Drive under a newly built train viaduct.

A TALE OF TWO TERMINALS

On November 1, 1958, the first link of the Glendale Freeway opened at a cost of $2.8 million. Vehicles could now safely bypass the tracks while zooming through Glassell Park. Far bigger ambitions awaited, of course. The 1958-1959 state budget allocated $14.2 million to nudge the freeway south to Silver Lake, which would include an interchange with the Golden State Freeway (eventual I-5). As this next 1.5-mile chunk got underway, officials began to acquire property farther west of the 5 in anticipation of the 2 Freeway reaching the 101.

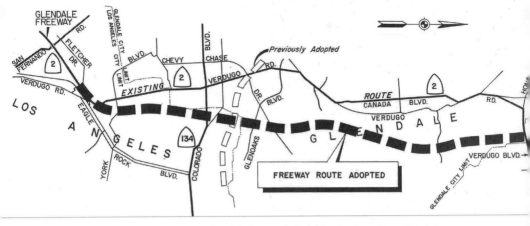

An early '60s imagining of the Glendale Freeway's first 6.3 miles from Glassell Park to Montrose. The route largely mirrors Verdugo Road, eventually taking that street's numerical designation—SR-2. (Note "North" is to the right.)

Meanwhile, north of the Eagle Rock Boulevard/Verdugo Road terminus, the goal was to complete a 6.5-mile link to the Foothill Freeway (I-210) by 1970. On June 14, 1962, one thousand people—dressed in their finest suits and dresses—gathered in the Glendale College auditorium to hear a presentation by the Division of Highways. The attendees, mostly from Glendale, were supportive of a freeway but anxious to know which alignment the agency was considering through their city. The Highway Commission presented four routes. Option One—also known as the Blue Line—was an easterly passage through the San Rafael Hills, away from the residential communities along Verdugo Road. Its displacement of 963 families in Glendale and Eagle Rock would be on the low end compared to the other three. It was also the cheapest route, at $48.7 million. Whenever speakers invoked the Blue Line at the lectern, the gallery interrupted with loud Pavlovian applause. Clearly, the crowd favorite.

Two weeks later, the state announced the adoption of the Blue Line. Homeowners were overjoyed, except for one documented sourpuss—a man named Alden C. Palmer. The seventy-year-old businessman's land near Glendale College was now condemned. He insisted that the state's projected construction figures were off, and formally asked it to reexamine more westerly routes that would, conveniently, spare his property. Palmer's protest was serious enough to halt work on the Blue Line and garner a response

FREEWAYTOPIA

from Glendale Mayor William Peters, who called him "the only contrary-minded person in Glendale." On August 23, 1962, the commission came back with their verdict: Glendale's loneliest man was uniformly dissed. The route would proceed as planned. Enthusiasts flocked to the lobby of Glendale City Hall to genuflect before an eight-foot scale model of the approved Glendale Freeway.

Contrasting the relative harmony of white, conservative Glendale, Angelenos on the south end of the freeway were bracing for an all-out war. Silver Lake and Echo Park were rich in ethnic and cultural diversity, with a proud artistic streak that hearkened back to the Silent Era, when the area was known as Edendale. The steep hills off Glendale Boulevard (née Allesandro Street) provided ready-made gags for the Keystone Kops and, later, Laurel and Hardy and the Three Stooges. Charlie Chaplin shot his first movie here, and Tom Mix rassled bad guys. In the 1930s and '40s, the area became a magnet for communists and free thinkers and drew a wave of immigrants, many of them Russians and Latinos.

By 1962, the region was a true melting pot. Throughout that year, highway officials held forums for those impacted by a proposed 2.7-mile extension of the Glendale Freeway to the Hollywood Freeway. The diagonal pathway would plow through mixed housing and hilly terrain between Glendale and Silver Lake Boulevards, jog west past Sunset Boulevard along Marathon Street and Melrose Avenue, and link up with the 101 at Vermont Avenue. (Reiterating the Hollywood Freeway chapter, northbound and southbound lanes near Vermont

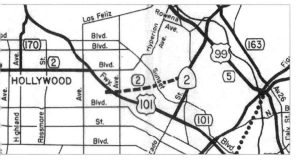

A map of the proposed extension of the Glendale Freeway (SR-2) to the Hollywood Freeway (U.S. 101).

were separated to accommodate an unbuilt junction with the Glendale/Beverly Hills Freeway.) Four alternate routes were presented to the masses. They ranged in cost from $24 million to $34 million and included, as a means of sparing more homes, plans for a double-decker in which inbound and outbound lanes were stacked upon one another.

On January 23, 1963, commissioners formally adopted the extension. The project would displace 4,650 people residing in 363 single-family homes and 768 multiple-unit dwellings. The state anticipated two to three years of additional property acquisitions, with construction beginning by the end of the decade. A *Los Angeles Times* headline—"NEW GLENDALE FREEWAY LINK WILL BEAR COSTLY PRICE TAG"—alluded to both the budgetary and human costs of a freeway through "well built-up" communities. Taking a page from the rousting of Boyle Heights residents for the Golden State Freeway, planners dismissed much of the proposed pathway as "blighted" or "a source of trouble for police." For low- and moderate-income residents along Melrose Avenue, Hoover Street, Heliotrope Drive, and other East Hollywood corridors, it was a dreaded case of déjà vu. Just ten years earlier, many of them had already relocated to make room for the Hollywood Freeway. Those in the foothills were equally outraged. "The Silver Lake area is a Swiss hamlet, surrounded by trees and mountains," said longtime resident Rosebelle Metcalf at one meeting. "You will crucify our love of aesthetics and beauty of Los Angeles on your commercial cross."

In a rare instance of a mixed-class community prevailing against freeway backers, the state decided to shelve the prospective 2.7-mile spur, limiting the freeway's southern hub to the Glendale Boulevard/Allesandro Street intersection. Residents were relieved, but there was no storybook ending. Instead, the saga of the 2 Freeway would devolve into a Lovecraftian nightmare that would extend well into the 1980s.

FREEWAY FATIGUE

With dreams of a connector to the 101 dashed—for now, at least—the Highway Commission turned its sights to SR-2's ascension past La Cañada Flintridge into the San Gabriel Mountains. Essentially, the agency proposed turning the entirety of the west-east Angeles Crest Highway into an eight-lane freeway. If you think that sounds ridiculous, consider what the state Legislature and L.A. County Board of Supervisors proposed: a 15.7-mile, $1 billion subterranean toll freeway from La Cañada Flintridge to the Antelope Valley. (Added bonus: backers said it could double as a "new escape route" in the event of a Russian attack.) Such an idea had been around since the mid-1940s, but in its new iteration, the tunnel would accommodate both vehicles and

The proposed tunnel route that would ostensibly link the SR-2 Freeway with the Antelope Valley Freeway (then U.S. 6).

high-speed trains. Amazingly, this proposal got rubber-stamped by the Senate Transportation Committee, though it died in the state Senate, which favored an overland route that would convert the north-south Angeles Forest Highway to the Angeles Forest Freeway.

By the mid-1960s—when it was thought that funding could only allow for one of the two San Gabriel Mountain thorough-fares—the Angeles Forest Freeway became the preferred route for extending the 2. Where the Angeles Crest Freeway would have ended in the desolate desert air of Victor Valley, the Angeles Forest Freeway would lead to the growing communities of Lancaster and Palmdale—the site of an international airport scheduled to open in 1968 (it was never built). From its junction with the 210 Freeway, the Angeles Forest Freeway would proceed as SR-2 for ten miles and then straddle the former SR-249 for thirty-two miles, eventually connecting with the Antelope Valley Freeway (SR-14).

Astonishing as it is to think about ravaging the alpine beauty of the San Gabriels, the freeway lobby still ruled the pre-environmental era. Both Angeles Crest and Angeles Forest were regarded as hazardous roads whose cliffside curves needed to be neutralized. This mindset was still apparent on L.A. freeway maps from 1965—dizzying tangles of solid and broken lines resembling the diagrams of a drunk football coach. Besides the jumble of complete or almost-complete freeways, there are proposed cuts through canyons, tunnels under mountains, and causeways over the ocean. But in a sign of things to come, one man said, "Enough's enough."

Frank Lanterman was a Republican assemblyman whose district included what was then La Cañada. You may remember him from the travails of the Foothill Freeway, which blazed through his hometown against his wishes. "I am still opposed to more freeways in the La Cañada area," he told the highway committee in 1968. "There isn't economic justification for the community to bear any more loss of land to the state for freeways." He had already conceded to the state's plan to widen the first mile of Angeles Crest Highway, which he felt was "ample to carry traffic north to the mountains." Though Frank didn't oppose a freeway through the San Gabriels in principle, his resistance to a La Cañada gateway—the only access point that made sense—created a glaring gap on planning maps. Lanterman's considerable influence led the California Legislature to scrub the La Cañada portion of an Angeles Crest/Forest Freeway. As a result, the state had no choice but to eliminate the entire mountain route from its maps. The 2 Freeway would go back to being a humble servant of the L.A. metropolis.

Lanterman's rejection of the SR-2 expansion coincided with public attitudes in the late '60s, which were shifting away from freeways in favor of cleaner transit options and more open space. Resistance also came from Washington. In 1967, $30 million in federal interstate funds to California were held up in the Senate. Though the Glendale Freeway was a state route, it endured collateral damage when the state prioritized major thoroughfares over "fringe" ones like the 2. The freeway would now blow

OFF-RAMP ↗

On June 15, 1978, the Assembly designated the segment of the 2 between the 210 and 134 as the Frank Lanterman Freeway. Only one person voiced a "no" vote—Assemblyman Lanterman, who was retiring. "I put up a hell of a fight against [freeways]," he griped. His colleagues enshrined him anyway, an acknowledgment of his twenty-eight years as an elected official who continually protected the foothill communities north of Glendale and helped expedite the completion of the Glendale Freeway once it was adopted.

Assemblyman Frank Lanterman, calling it a career in 1978.

past its 1970 completion date. "GLENDALE MAY BE 10 YEARS AWAY," the *L.A. Times* warned ominously.

As the years dragged on, Glendale's succession of mayors made annual treks to Sacramento to try to loosen up purse strings. Without the freeway, said Mayor Kenneth R. Stephen, "we have on Verdugo Road the worst community traffic bottleneck in the state," an allusion to the 65,000 vehicles that passed by Glendale College every day. Glendale itself spiked to the tenth largest city in California, with 137,000 people. The impending freeway drove much of this growth, leading to more business development and housing. "Glendale is unique for a large city in that it is without slums," crowed a real estate blurb in a local newspaper.

When the freeway-adjacent Alpha Terrace condominium complex opened, a print ad promised nothing short of career advancement for new owners. It included a mock testimonial written by an owner's housewife, who was prepping for an important dinner guest—her husband's boss. "I was so nervous, I wanted everything to be just right..." she confided in the ad copy. Turns out, she and her husband "Harry" had nothing to worry about: "'Why Harry,' the boss said, 'you live like a top executive. Just look at this hillside setting and these views from your private balcony ... it's incredible!' I overheard him say to his wife, 'That Harry, he is definitely executive material.'" The ad ends with the boss's wife selecting carpet and draperies for their own Alpha Terrace condo "before we had even finished dessert."

In the early '70s, Caltrans priced out the final costs of the Glendale Freeway's missing links: $16.4 million for the 3.3 miles between Eagle Rock Boulevard/Verdugo Road and the Ventura Freeway; $13.8 million for the 3 miles between the Ventura Freeway and Fern Lane (just south of the 210); and $19.2 million for the Glendale-Foothill Freeway Interchange, which included the final mile of the Glendale Freeway. True to the *Times'* projections, work proceeded at a snail's pace. State funds for the Glendale Freeway were not expected to be available until 1974-1975.

Easterly view in early 1974 of the unfinished Glendale Freeway passing over the future Ventura Freeway extension, which is still in the grading stage.

On a national level, the energy crisis that delayed other Southland freeways also ensnarled the Glendale. Even tractors weren't immune to gas rationing. Kirst Construction—one of the freeway's contractors—had to cut back to 20,000 gallons of diesel fuel a month. They needed 225,000 gallons to stay on schedule.

By the fall of 1975, it was looking like the Glendale Freeway—and its neighboring freeway, the Foothill—would never be completed. The Glendale-Foothill Interchange had five towering ramps leading nowhere, eerily resembling the severed post-"Big One" freeway bridge in *Isle of California,* a famous 1972 mural on a building off Santa Monica Boulevard (aka State Route 2, coincidentally enough). The Glendale-Ventura Interchange also had ramps eternally suspended in mid-air. Long a supporter of freeways, the *L.A. Times* penned their obituary: "Clearly, the freewheeling, freeway-building days of the early 1960s, when every mile of freshly paved freeway was considered a blessing, are gone."

And then, like a phoenix rising from the ashes, good news arrived in the state's 1976 budget. The Glendale Freeway was allocated finishing funds. In June of 1976, the final chunk between the Golden State and Ventura Freeways opened to traffic. The milestone, however, was marked by tired cynicism. When asked who would christen the freeway, twenty-five-year Caltrans veteran John Meenan simply shrugged: "The first guy driving by . . . some nameless soul who just happened to be there." Meanwhile, the *Times* continued its lament of a past era. "Freeways have become political liabilities," the paper reported. "The day of ribbon-cuttings with politicians wielding huge gold-sequined scissors, the days of champagne and speeches is long gone. It is not sound politics, in an era of environmental concern, to be pro-freeway." The parade-raining didn't stop Meenan or his crew from savoring the moment. He and the contractors cracked open

END OF THE LINE—John Meenan, Caltrans resident engineer, stands near unfinished ramp connected with eastbound traffic.
Times photo by Leo Mark

Several years into stop-and-go construction, a newspaper caught Caltrans's John Meenan under a half-built ramp of the Glendale-Ventura Interchange with a look that says, "Are we ever gonna finish this thing?"

celebratory beers on a bridge abutment. Then Meenan took a test-drive on the fresh concrete at speeds over eighty miles per hour, "to test for bumps," he winked.

The freeway's final four miles between the Ventura and Foothill Freeways opened in March of 1978. This was one event that couldn't be ignored. The city of Glendale hosted a dedication that recalled the good ole days of antique cars and a carnival atmosphere. Dignitaries included Los Angeles Mayor Tom Bradley and Glendale Mayor Carroll Parcher, who effused, "People have been talking about this freeway for twenty years, and they're eager to use it!" As for the Glendale Freeway's extension to the Hollywood Freeway? A Caltrans official rebuffed rumors that the agency was planning to resurrect their plans. "This is it. There no longer is any money," he said, as estimates had inflated to $45 million. "Sure, there will be a few more miles built, just to fill in the gaps, but our freeway system here . . . well, it's seen its heyday."

A PHANTOM MENACE

Though the 2.7-mile link to the Hollywood Freeway was killed by the Legislature in the early '60s, it loomed as a spectral presence

OFF-RAMP ↗

The Glendale Freeway through the San Rafael Hills was the first to utilize "stepped-sloped landscaping." Facing criticism for its former practice of sheer cuts into the mountainside, Caltrans's landscapers carved a series of steps into the cliff face and planted them with native shrubs and vegetation. Besides restoring the hills' natural look, the shelves reduced erosion and acted as seed cultivators when loosened soil fell onto each step.

CALIFORNIA NATIVES

EASTERN UNITED STATES

MEDITERRANEAN REGION

SOUTH AMERICA

CHINA AND EAST ASIA

AUSTRALIA AND NEW ZEALAND

SOUTH AFRICA

OTHER

Origins of plants by geographic region for freeway landscaping in the 1960s. Over thirty California native trees and shrubs were used, although the Mediterranean region contributed two hearty staples: oleander and Algerian ivy.

over East Hollywood and Silver Lake. Renters forced to sell their previously owned homes to Caltrans found that they could not buy them back. A state of limbo lingered well into the 1970s, taking a toll on all who lived or worked in the former right-of-way zone. Businesses deteriorated and property values declined. Homeowners didn't know whether to sell their properties or hold on to them. "It's like a black cloud," said one resident. Similar perpetual angst would hover over the proposed 710 Freeway corridor in South Pasadena. That controversial link was rescinded by the federal government in 2003, but Caltrans didn't start selling the properties it acquired until well over a decade later. So what exactly was going on here?

One culprit was the Beverly Hills Freeway, lurking about like an unwelcome party guest.

Holding signs like "CALTRANS IS SELLING US OUT!", picketers march in front of Caltrans's L.A. headquarters on September 7, 1977. At issue were individuals who lost their homes to the proposed Glendale Freeway extension, or whose fates were made uncertain because of it.

Despite fierce opposition from Westside and Mid-City residents, the route stubbornly clung to annual planning maps as a dotted line. The other instigator was Ronald Reagan. For three straight years, the then-governor of California vetoed legislative bills that would have expunged that freeway. Highway officials found themselves in a pickle: if the Beverly Hills Freeway moved forward, they would have no choice but to revive the Glendale Freeway link to the Hollywood-Beverly Hills Freeway Interchange. Clearly, an uninterrupted flow for the 2 Freeway would be the only way to justify the Beverly Hills Freeway's $315 million price tag ($7 million had already been paid out for rights-of-way). Seeking to avoid having to buy back homes it sold at a much higher price, Caltrans held on to its role as a reluctant landlord.

A breakthrough occurred in 1975. With Reagan termed out as governor, the Legislature deleted the unbuilt gap from the freeway system. Juan Vega—a tenant on Mariposa Avenue in East Hollywood—was among those excited to repurchase his former home from Caltrans, only to find himself tripped up in red tape. Under existing laws, Caltrans held that homes should only be sold at public auction. Vega formed a coalition called the Route 2 Tenants' Association, which petitioned Governor Jerry

A Caltrans sketch of the Beverly Hills Freeway, 1971. It envisioned the relocation of train tracks running down Santa Monica Boulevard to the freeway's center median. In addition to this trenched route, cut-and cover (tunneled) and elevated routes were also considered.

Brown and enlisted the Department of Housing and Community Development to allow former homeowners first rights of refusal on repurchasing homes. Siding with the tenants, the California Attorney General authorized a moratorium on public sales to allow Caltrans time to proceed with direct sales to tenants. More delays ensued as Caltrans started the long process of assessing the properties it had acquired to prepare fair market values—all told, 123 single-family houses and 421 apartments affecting 1,500 people.

Finally, on April 18, 1978—one month after the Glendale Freeway opened—Caltrans announced that it would sell off its previously condemned properties. However, when their appraisals were revealed, prices had soared some 50 percent during a real-estate boom. In a cruel bit of irony, the bump was partially attributable to the news of the now-dead freeway spur. Vega suddenly found himself unable to meet his former home's market price. "It was so frustrating," he recalled. "You couldn't get a straight answer out of [Caltrans] the whole time. They just made you sweat and sweat. It was as if they didn't really care what happened to us."

To remedy this problem, the state passed legislation that enabled tenants to buy back their homes at 1975 market values. It was a novel solution, but it was not met with universal support. Longtime homeowners in the area feared the lowering of their own property values. Pro-business groups argued that bypassing the traditional auction process was not just an illegal use of gas tax highway funds, it was also unconstitutional. The threat of litigation caused further delays, but others saw something more insidious at play. "It's disgraceful," said state Senator David Roberti, who represented the district. "Caltrans is acting like a sore loser, letting people suffer, letting the neighborhood deteriorate just because its freeway was killed and its plan to auction off the condemned properties to speculators was overruled." Caltrans defended their deliberations. Selling homes below market value was an unprecedented move that required proper legal mechanisms.

In 1979, a legal safety net arrived with the passage of a Senate bill that created a "mitigated negative declaration," an instrument that allowed Caltrans to write down the cost of its properties so that lower-income tenants could afford to buy them based on formulas commiserate with their incomes. This was a lifesaver for tenants looking to re-own their units. One study showed that the average yearly income in the area was $12,380. Another study revealed that only six out of 123 families in the affected zone could afford a single-family house at market value. Now, almost everyone had skin in the game. The only real loser was Caltrans. "When you spend $35,000 to rehabilitate a house and sell it for $22,000 or $23,000, you aren't going to make much money," a Caltrans "excess land" agent wryly observed. Indeed, the agency came up well short in recouping its total acquisition costs of $6.4 million.

But the benefit to the Silver Lake-area community was immeasurable. Though it would take several more years to work out the kinks—most of the home sales weren't finalized until 1982—the program reinforced community stability. "A nightmare has turned into a dream," enthused Bill Angel, who bought back the home he lost in 1968. Josephine Pellulo had also been paying rent to Caltrans for fourteen years, fearing eviction every year. "This is the only home I have known since coming from Naples in 1949," she said. "My children were born and raised here. All

With echoes that still resonate today, apartment tenants march in front of Mayor Tom Bradley's house in 1982 to call for rent control. The ordeal of the Glendale Freeway underscored the growing sense that L.A. regulations overly favored landlords, of which Caltrans technically was one.

my memories are here. And this is where I want to die." Even apartment-dwellers enjoyed the privilege of ownership. The Senate bill called for multiple family units to be organized into limited-equity cooperatives, allowing renters to acquire their units at "affordable prices." The program created the largest cooperative housing project in the West, and became a model for other lower-income housing plans in Los Angeles.

OFF-RAMP ↗

In 1986, fears of a freeway extension through Silver Lake were revived when the county and the Central City Association proposed a route that would straddle Alvarado Street, intersect the Hollywood Freeway, and hook up with the Harbor Freeway at Slauson Avenue. "It's absurd," said Councilman Michael Woo. "The fact that they are even talking about it is laughable." Pegging costs for the project at $800 million, Caltrans shot down this "downtown bypass" freeway.

The Vegas were among the 500 families in 1982 who became owners anew. Juan Vega knew just how to celebrate their victory against the state. He suggested that the Route 2 Tenants' Association change its name to the Route 2 Tenants' and *Homeowners'* Association. "It seems appropriate," he said.

At long last, the idea of wedding the 2 to the 101 appears to be dead as a doornail. But Caltrans has never resolved the issue of the unsightly speedway that is Glendale Freeway's southern terminus. In 2001, engineers envisioned a $14 million reconfiguration. More recently, state grants have been proffered to fix the problem. But no one has been able to articulate a vision that maintains traffic flow while enhancing pedestrian accessibility and safety—two needs that lie at odds with each other. At a 2012 community workshop, some residents suggested replacing the one-mile freeway stub off the 5 with parks and housing, a West Coast version of New York's High Line. At the very least, critics argue that Caltrans should landscape the slab. Some even theorize that the lack of permanent action is purposeful. After all, wouldn't that be admitting defeat?

PREEMINENT DOMAIN

I'll admit to a biased attachment to the Glendale Freeway. For years, I stared at its ugly Silver Lake crossover ramp from a dugout bench at the Tommy Lasorda Field of Dreams, where I coached my son's youth baseball teams through the 2010s. When the field opened in the late '80s, parents worried that vehicles might fly off the ramp and crash through the thirty-foot chain link fence—the only thing separating the freeway from right field. (Cars launching off a freeway onto a park named after a legendary Dodger . . . what could be more L.A.?) Truth be told, the greater threat came from the league's budding sluggers, several of whom launched moon shots over the fence and onto the freeway. One benefit to

OFF-RAMP ↗

Throughout much of 1991, an unknown party deposited 350 tons of dirt onto the Silver Lake freeway spur over several nights. Caltrans officials could only scratch their heads at the makeshift mountain, which spread hundreds of feet wide. James Bonar, president of the Silver Lake Residents Association, wondered if the phantom dumper was, in fact, Caltrans, "exercising their resentment against the Silver Lake neighborhood for rejecting the extension of that highway through our community." The agency eventually hauled the dirt away and used it as backfill for work on the Glendale-Ventura Freeway Interchange.

The southern terminus of the Glendale Freeway. As motorists descend onto Silver Lake surface streets, they must also be mindful of baseballs smacking their windshields, thanks to a short right field on the other side of the tall fence.

all: local leaders believe the much-needed neighborhood ball field effectively blocks Caltrans from considering future expansion of the 2.

Farther north—between the 5 and 210 Freeways—the 2 remains a seductive strip. Besides the Arroyo Seco and Marina, no freeway in Los Angeles generates more feels than the Glendale. For many, it's the views, the sparse traffic, the occasional sightings of deer grazing on the hills west of Descanso Gardens. Even in a KPCC poll, listeners found the Glendale Freeway to be "the most tolerable." And how many other freeways inspire an open love letter? In her EastsiderLA essay, "In Love With The 2: An ode to a Northeast L.A. freeway," Brenda Rees makes the case that the Glendale is the rare freeway "that promises—and DE-LIVERS—an exhilarating driving experience with matching natural scenery."

Well, it didn't exactly deliver on its original promise to connect the sandy climes of Santa Monica to the snowy slopes of the San Gabriels. And thank goodness for that. Unburdened by its loftier visions, the Glendale Freeway was left free to be a wide-open canvas onto which Angelenos can paint their own super-highway nirvana.

A picture-perfect portrait of the Glendale Freeway, with the Downtown skyline rising like a mythical City on a Hill.

Chapter 9

THE SANTA MONICA FREEWAY
Interstate 10 / State Route 1
(1961–1966)

If the Arroyo Seco is the grand dame of L.A. freeways, the Santa Monica is a post-modern descendant. Coming of age in the 1960s and '70s, it was a lab rat for experimentation and far-out ideas, man, forging a cosmic journey to the coast during the height of beach culture and surf rock.

In some ways, the 10 Freeway's first electronic "test" message in 1973 is presciently on point.

The Santa Monica Freeway pioneered electronic message boards. When they weren't flashing life-affirming bromides to Steve Martin in *L.A. Story*, these lighted signs were imploring commuters to check out the 10's newfangled Diamond Lane ("Car-pool is bliss for lad or miss!"). And in a male-dominated field, the freeway bore the imprint of two influential women—one a visionary engineer, the other a transportation big shot.

By the mid-1970s, however, the Santa Monica underwent the freeway equivalent of middle school. The Diamond Lane was its undoing. The roadway was cursed, vandalized, the butt of nationwide jokes. The woman who managed the doomed carpool lane endured sexist taunts and death threats. The freeway was also disparaged as a line of demarcation in a racially segregated city. Through it all, traffic swelled like a cancer. The only artery between L.A.'s Westside and Eastside, it was the freeway you couldn't live with...or without. Its rise in status was ready-made for a coffee mug: World's Busiest Freeway.

And then, after three decades of thankless servitude...it collapsed. A major earthquake took out two of its bridges. The flashy upstart now looked old and frail. That's when everyone rallied around the Santa Monica Freeway, from traffic-weary haters to President Bill Clinton. The Santa Monica was deemed the MVP among L.A. freeways—annoying, sure, but absolutely essential. Crews worked twenty-four-hour shifts to get it back online in a miraculous eighty-six days. Resurrection complete, the freeway has gone on to live a relatively unremarkable existence in the years since.

But make no mistake, its voyage from inception to completion had more ups and downs than the old Whirlwind Dipper roller coaster at Santa Monica Pier. As one of the freeway's message boards might say: "Strap in, it's going to be a wild ride."

Traveling almost 2,500 miles across America, I-10 ends with a cinematic-type sensation through the McClure Tunnel.

Anyone who has driven the Santa Monica Freeway toward the beach is familiar with the McClure Tunnel. It is a transformative thing, really, this four-lane portal to Pacific Coast Highway. As you cruise through the freeway's final trench—the western extremity of the 2,460-mile Interstate 10—its drab curved walls soon reflect blinding sunlight. It reminds me of the old Ice Tunnel at Universal Studios, where shimmering ice sheets swirled around the tram as if to suggest passage to another dimension. Seconds later, your car is spit out onto PCH. This is your cue to roll down your window, breathe in the salty air, and take in the view. Monochrome beige has been replaced by a Technicolor kaleidoscope of brilliant blue sky, glistening aquamarine seawater, white sand, and rows of waving palm fronds, while, on your right, patchy green vegetation cascades down the ruddy cliffs of Palisades Park. The change is so extreme, the best comparison might be when *The Wizard of Oz* switches from black-and-white to color to represent Dorothy's journey into a mythical land.

After serving trains for decades, then-Palisades Tunnel was converted to a car tunnel starting in 1934. Palisades Park's cliff is on the left, Pacific Coast Highway on the right.

The Santa Monica Freeway's I-10 designation ends at Lincoln Boulevard, which is part of State Route 1. From Lincoln to the tunnel, the freeway's last half-mile is technically SR-1. Long-term plans were to have the freeway link with the Pacific Coast Freeway. One segment of this never-built coastal thoroughfare that was seriously considered was a six-mile, four-lane causeway across Santa Monica Bay, bridging Santa Monica and Malibu over a series of man-made islands. *Gunsmoke* star James Arness was among those who helped shoot down the cockamamie scheme.

As seen in the A+D Museum's "Never Built" exhibit, the "Sunset Seaway" would've hopscotched islands excavated from the Santa Monica Mountains to the tune of 120 million cubic yards of earth.

So who was this McClure fellow, and do we have him to thank for this magical tunnel? Well, no. And yes.

A tunnel running under Ocean Avenue dates back to 1886, built by Southern Pacific Railroad to connect with its Long Wharf cargo pier. (An 1898 film by Thomas Edison—*Going Through the Tunnel*—shows a train passing through and is viewable on YouTube.) Engineers with the Works Progress Administration reconfigured the tunnel for cars in 1936. Other than the addition of a few safety features, it was integrated as-is into the Santa Monica Freeway upon its completion in 1966. Three years later, it was dedicated to Robert E. McClure—a popular local figure who marshaled his power and influence to make sure this transcontinental yellow brick road ended in his Oz.

A UNIFYING DIVIDER

Outside of Santa Monica, however, McClure was anything but revered. He was already in a position to help shape public policy as

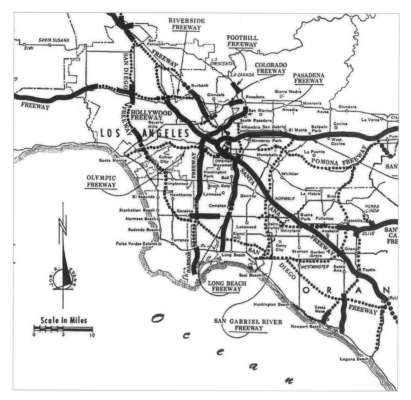

January 1, 1957: After securing federal funds, the Department of Public Works rolls out a map that shows the "Olympic Freeway" budgeted from Downtown to the sea. The route would become known as the Santa Monica Freeway later that year.

the long-time editor of the *Santa Monica Outlook*. Then, in January 1954, he was appointed to the state highway commission for a five-year term. McClure was your classic overachiever—impatient, imperious, a "results" guy. As such, he wasn't keen on public hearings; all they did was "hold up freeway action throughout the state." Nonetheless, Angelenos were granted their first forum on this new freeway in early '54. It was to be called the Olympic Freeway, because it would parallel Olympic Boulevard from East L.A. to Santa Monica.

At the meeting, California Highway Commissioners mapped out possible routes. Some would cause more disruption than others, but initial projections calculated that a ten-mile stretch

In 1946, picketers protest restrictive covenants at the house of Henry Law, a Black Angeleno, after he was told to move out of his predominantly white neighborhood. The U.S. Supreme Court ruled against discriminative zoning in 1948, but the practice continued through workarounds.

from Interstate 5 to La Cienega Boulevard would destroy 2,000 structures and displace 10,000 people—many of them Black families who would have difficulty relocating due to restrictive housing laws. Representatives for these shunted households called on County Supervisor Burton Chace to work with the County Housing Authority and help them find new homes.

But options were limited—certainly in the suburbs, where housing discrimination was rampant in these times before the Fair Housing Act. Decades later, returns on homes in the burbs have paid off like a Lucky Seven jackpot, allowing generations of families to accumulate wealth. All the while, Blacks—many of whom had moved to

Gone with the Wind actress Hattie McDaniel, the first African American performer to win an Oscar.

Los Angeles for work during the Second Great Migration of the 1940s—were deprived of home equity. African American housing activist Pastor Stephen "Cue" Jn-Marie, who lived in the Crenshaw district for years, viewed the Santa Monica Freeway as a kind of unifying dividing line. "Freeways connect us in such a way that the 'haves' have, and the 'have-nots' still don't have," he said.

It wasn't just working-class Blacks who were affected. The historic West Adams area used to be called Sugar Hill, home to some of L.A.'s most prominent African Americans. Actress Hattie McDaniel was a pillar of the community. When she wasn't holding community workshops, she was throwing epic parties that drew Black America's most influential business leaders and entertainers. Within a decade, dozens of Victorian and Craftsman homes in Sugar Hill would be wiped out; those whose homes were spared chose to move away. "The road could have been built without cutting through the so-called Sugar Hill section," the *Los Angeles Sentinel* asserted. "However, in order to miss Sugar Hill, it was 'said' that the route would have to cut through [the] fraternity and sorority row area around USC." The Greek housing rows still stand.

Farther west, a small but vocal group of predominantly white homeowners pushed back against the freeway's seven-mile extension to Santa Monica. At a 1955 town hall, a West L.A. doctor called the route "both logically and morally wrong, especially from a health standpoint." Residents of Cheviot Hills and Rancho Park pointed out that the high-speed road would pass within a few hundred yards of several schools. The owners of the Vista Del Mar Orphanage—which would lose four of its twenty-one acres—warned that the freeway's noise and proximity would cause severe mental distress among its already emotionally fraught children. But even the cries of an orphanage could not dissuade Robert E. McClure. "I am very proud that I am singled out by the freeway obstructionists as the man who advocated a freeway for the western part of Los Angeles County," he said.

Like it or not, in other words, the route was fixed.

But the Westsiders weren't done. At a subsequent meeting, dissenters circulated a petition arguing that McClure, as a civic leader in Santa Monica, should step down from the highway commission due to a "personal interest in [the freeway's] routing. Assemblyman Thomas Rees (D-Beverly Hills) was among those who felt McClure's position created a conflict of interest. "If a member of the highway commission is to be disqualified because he advocated freeways," McClure shot back, "all of us would be disqualified." Painting himself as a martyr, he grumbled that he was actually losing money—his expenses exceeded his measly state salary—for the "thankless" task of agitating for the greater good. The project had broad public support, and it fell on McClure to make sure municipal, state, and federal agencies stuck by their promises. "I have worked for nearly ten years to get us a freeway," he said. "And we are still waiting for one."

Robert McClure's hard line wasn't borne out of obstinacy. As a newspaper guy, he had a good read on public sentiment. The Santa Monica Freeway—its official name as of 1957—was perhaps the most well-received project during the heyday of freeway-building. By the late '50s, the Westside accounted for 15 percent of Los Angeles's population, greater than that of St. Paul, Minnesota. A link to the sea was first bandied about in the early '40s, and was considered long overdue. The main source of its delay was the Eisenhower Administration, which needed time to draft its interstate network before settling on the Santa Monica Freeway as the last leg of Interstate 10. With that designation, the freeway now qualified for federal highway funds for both

> **OFF-RAMP** ↗
>
> Though the California Highway Commission changed freeway names to reflect their destinations in 1957, it was not an ironclad rule, as the Century and Golden State Freeways demonstrate. And, of course, while the Arroyo Seco Parkway became the Pasadena Freeway, it switched back to the Arroyo Seco Parkway handle in 2010.

rights-of-way and construction. No one was more thrilled than Santa Monica's grand poobah. "The best thing for Santa Monica was to have this place designated as the western terminal for the national highway system," McClure raved. For a driver hopping on I-10 from Jacksonville, Florida, that meant "the name Santa Monica would be on all highway signs leading to the coast."

The government's involvement, once slow to come around, now created a sense of urgency. "Use it or lose it" was the usual credo when it came to spending federal money—here, 90 percent of the freeway's $202 million budget. Right-of-way agents

Construction of the Santa Monica Freeway southwest of Downtown L.A., with bridges over the Harbor Freeway built in advance.

prowled neighborhoods from Interstate 5 to West L.A., knocking on doors in the freeway's path. One agent characterized homeowners as so welcoming, one almost imagined them offering these bearers of bad news plates of warm cookies. "Well, we've been expecting you!" most of them allegedly told the agent. Only two out of every hundred owners took the state to court to contest their evictions or wrangle a better price. (This was about on par; only 3 percent of eminent domain cases made it to trial.) In all, about 15,000 people lost their homes to the Santa Monica Freeway.

A 1959 map displaying the "loop" of freeways around Downtown to help disperse traffic. Reflecting the rigid priorities of the era, district engineer Lyman R. Gillis described the loop as "the best freeway location from a traffic service standpoint at the lowest possible cost."

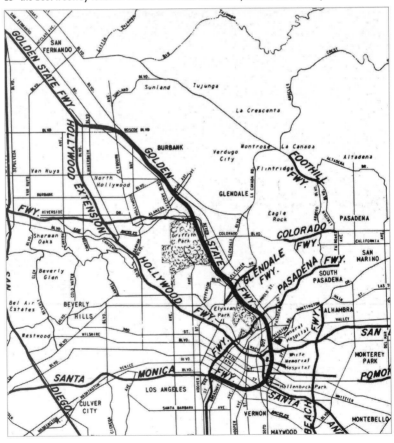

FREEWAYTOPIA

The freeway's groundbreaking was a viaduct over the L.A. River in 1957. "Not since Roman chariots armed with hooks and scythes raced up and down Italian highways will there be so many wheels turning over so many lanes," read the lede in the *Los Angeles Times* that year. The newspaper, as expected, was gung-ho on another freeway, trotting out facts, figures, and neato maps that emphasized economic benefits and traffic alleviation. By creating an "eastern loop" that connected with the Santa Ana, Golden State, and San Bernardino Freeways, the Santa Monica Freeway would siphon traffic away from the Four Level and central Downtown. Answering charges by urban planners that freeways are often outdated by the time they open, the *Times* reassured its readers with a quote from Edward T. Telford, the engineer in charge of District 7 (L.A.'s freeway district). "The standards we're applying to them now will make them retain their capacity to serve traffic permanently," he pledged. "We're designing the freeway for operating speeds of approximately sixty miles per hour. If at the end of fifteen years, people want to operate at a hundred miles per hour, then the freeway will be obsolete." Needless to say, Telford's standards did not hold up, but then, neither did his logic.

With the progression of surveys and land acquisitions, the construction timeline was broken into three stages: Santa Ana-Golden State Freeway (I-5) to the Harbor Freeway (then SR-11); Harbor Freeway to La Cienega Boulevard; and La Cienega to Santa Monica. From an engineering standpoint, the first stage was the most challenging. This 3.6-mile segment required $8,400,000 worth of bridges over a heavily industrial zone that included the Los Angeles River, numerous warehouses, and the Santa Fe Railroad yard. From there, it would continue on concrete stilts—2,125, to be exact—over L.A.'s wholesale district (also deemed too important to raze) before crossing over the Harbor Freeway.

On December 4, 1961, the first 1.1 miles of freeway opened to

Going up! The Santa Monica Freeway's elevated portion under construction west of the Golden State Freeway, 1961. A warren of underground utility conduits due to heavy industry was a deterrent against a grade-level freeway here.

cars between Hooper Avenue and Main Street. The ribbon-cutting was followed by an even bigger affair three months later to celebrate the Santa Monica-Harbor Freeway Interchange and another 1.1 miles of virgin roadway. Four hundred spectators gathered on the Maple Avenue on-ramp to watch Governor Pat Brown and a parade of Freeway Queens cut the ribbon with oversize scissors. The ribbon was attached to balloons, which were supposed to flutter skyward after they were cut loose. Sadly, someone forgot to bring the helium. The balloons drooped and skidded on the hot pavement.

This was a mere trifle compared to a far bigger faux pas that lurked in the future—one that would make the Santa Monica Freeway the laughingstock of the nation.

> **OFF-RAMP** ↗
>
> During construction of the Western Avenue ramps, someone stole a three-ton asphalt roller from the freeway in the middle of the night—a $10,000 heist. Police were never able to find the thief, and determined that several Flat Stanley sightings in the neighborhood were merely coincidental.

FREEWAYTOPIA

FREEWAY-A-GO-GO

But first, in the immediate future, there were only triumphant milestones: The groundbreaking in Santa Monica in 1963. An extension to La Cienega in October of '64. Bundy in January of '65. Crews were pouring concrete in record time at a record cost. Justifying its $12 million-a-mile price—the most expensive freeway ever built—spokesmen with short memories called the motorway to the beach "the most anticipated event in the twenty-five-year history of the freeway system." Previously referred to as "Freeway Orphans," 500,000 Westsiders now had a lifetime free pass to L.A.'s ultra-cool commuting club. Dedications for each new off-ramp outdid previous ones. Forget street-level ribbon-cuttings and flaccid party balloons; for the opening of the Robertson/National bridge, officials hired a Goodyear Blimp with enough dang helium to fill 60,000 balloons. The airship swooped 150 feet above the eastbound lanes with a roped tire dangling from its gondola. The tire tore through the ribbon, clearing the way for a motorcade to the strains of a Dixieland

Surrounded by swimsuit models from the International Auto Show, Robert E. McClure, left, is all smiles as Governor Pat Brown honors him at the freeway's La Cienega extension in 1964. Peeking over McClure's shoulder is future Mayor Tom Bradley, then a city councilman.

A partial aerial view of the Santa Monica-San Diego Freeway Interchange. The large footprint allowed for sweeping, 75-foot-tall flyovers engineered for freeway speeds.

band (because . . . why not?). Governor Pat Brown dedicated the freeway to Robert E. McClure for his efforts in bringing the "road paved with gold" to SoCal's sandy shores. McClure would bag his biggest trophy—the McClure Tunnel—five years later.

The western front's crowning achievement was the 10 Freeway's connection to the 405 in November 1964. The $25 million, three-level interchange established several benchmarks. It featured the longest single connector of any junction—*LA Weekly* readers once voted the soaring transition from the 10 East to the 405 North as the city's best freeway ramp—and was the first to allow for speeds of fifty-five miles per hour on all ramps. It was also one of the first designed with computer

OFF-RAMP ⬈

The Santa Monica Freeway's other innovations included: the use of white letters on green signs (the old white-on-black ones were switched out on other freeways in short order); Accident Investigation Sites (AIS), which are turnouts where the CHP can clear accidents to reduce rubbernecking (though I have yet to witness them in action); and extensive stretches of auxiliary lanes—sometimes called collector lanes—to serve local traffic (and impatient rogues looking to dart around congestion).

Marilyn Reece, early in her 35-year career as a civil engineer for the state. "By far the smartest person I've ever known," recalled her daughter, Anne Bartolotti.

assistance, and inspired future Caltrans engineer Arturo Salazar when he drafted the beautifully fluid Santa Ana-Costa Mesa Freeway Interchange. Historian Reyner Banham described the Santa Monica-San Diego nexus as "a work of art, both as a pattern on the map, as a monument against the sky." The project was notable for one other reason: it was the first interchange designed by a woman.

Born on a farm in North Dakota, Marilyn Jorgenson Reece moved to California in her twenties. She was the first woman in the state to become a fully licensed civil engineer, and her timing was fortuitous. The Division of Highways needed a slew of engineers for their freeway program, but qualified men were in short supply after World War II. Reece was hired as the agency's first female engineer and quickly rose through the ranks of its 16,000 employees. The general public's introduction to her—and another woman engineer, Carol Schumaker—came under this of-the-era *L.A. Times* headline: "FREEWAY BUILDERS ARE WEEKEND HOUSEWIVES."

Reece with fellow engineer Carol Schumaker, left, who designed an interchange in Orange County and rose to the highest-ranking female engineer in the Division of Highways.

THE SANTA MONICA FREEWAY

In the article, Reece was lauded for shattering the concrete ceiling of a male-dominated industry, but the female writer was most impressed by her ability to hold down the home front as a mom to two toddlers and wife of a fellow engineer. When she wasn't working, this blond dish from Whittier looked forward to cooking, cleaning, and herding the kids—just like moms everywhere! "I go to the market and wash and feel just like the girl who stays home all the time," Reece said, almost apologetically. "You couldn't tell the difference." Any mentions of her work life were girded by subtext that women like Reece represented no threat in a *Mad Men* world. She didn't go out and "boss a construction job or move heavy equipment about." Visits to the project site were "just to see how it's going." When Reece mentioned she didn't get bonuses "or anything like that," the writer insinuated that women engineers like her would probably prefer to be rewarded in the looks department anyway, "monumentally satisfied [if their] accomplishments were measured in inches or weighed in pounds."

There was, of course, far more to Marilyn Reece than lazy clichés. Her younger daughter, Anne Bartolotti, attested that she "cursed like a sailor." Burly contractors often underestimated this petite woman, who showed up to the interchange's

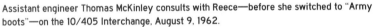

Assistant engineer Thomas McKinley consults with Reece—before she switched to "Army boots"—on the 10/405 Interchange, August 9, 1962.

OFF-RAMP ↗

Marilyn Reece's elder daughter, Kirsten Stahl, also worked for Caltrans's District 7, pioneering quick-drying techniques for freeway pavement jobs in metro Los Angeles.

construction site in Bullocks Wilshire suits and heels. But if a contractor was skimping on work details, "she would make their life a living hell," and her superiors would back her up. In all, she reworked thirty different design plans, many dealing with new specifications to avoid a church, an apartment building, a sanitarium. Bartolotti liked to say she hung out on-site as much as her mom by virtue of the fact that Reece was pregnant with her much of the time. Later in her career, Reece designed the bridge that gapped the "missing link" on the Foothill Freeway. A young Bartolotti sometimes accompanied Mom and skateboarded along the unfinished lanes. By this point, Reece had switched out her high heels for boots, which were better suited for the field. "I remember the insult at school back then was, 'Your mother wears Army boots,'" Bartolotti recalled. "And I'm like, 'How'd you know?'"

Construction west of the Santa Monica-San Diego Interchange continued through the '60s at a rapid clip, which thrilled everyone but the people pushed out of its way. In Santa Monica, 550 families were displaced, including Black residents living in a low-rent corridor along the freeway's contours. Middle-class Black households living by the beach were also forced to move. As with

Protestors at the Board of Realtors office, 1964. After California passed the Fair Housing Act in 1963, real estate professionals spearheaded its repeal, leading to continued discrimination. The act was restored in 1966 by the California Supreme Court.

THE SANTA MONICA FREEWAY

the mid-city portion, many were concerned about finding new residences, and their fears were well-founded. The Fair Housing Council conducted door-to-door and phone surveys that revealed most landlords and homeowners were reluctant to rent or sell to African Americans, even those with "substantial incomes." By one count, only one out of twenty-seven apartment managers said he'd definitely rent to a Black person.

When Black families did relocate to heavily white neighborhoods, the Fair Housing Council often sent escorts to shield them from threats of violence. Four years earlier, when an African American family moved into the white Sunset Park area south of Pico Boulevard, they were greeted with two Molotov cocktails thrown on their front porch. This scenario may seem incongruous with present-day Santa Monica, but in the early '60s it was still a conservative town with a history of hostility toward minorities. In 1920, a Black man named Arthur Valentine and his family were ordered off of a "whites only" beach by the police. When Valentine resisted, he was beaten and shot. Two years later, homeowners formed the Santa Monica Bay Protection League, whose not-so-hidden agenda was to keep Blacks out of the pricey coastal region. As a newspaper headline bluntly put it, "Settlement of Negroes Is Opposed."

On January 5, 1966, the final stretch of the Santa Monica Freeway opened between Bundy Drive and Pacific Coast Highway. Santa Monica Beach and Pier would soon be flooded with a surge in foot traffic, and Santa Monica commuters would no longer have to endure 119 traffic lights along Wilshire Boulevard to get Downtown. The total cost for

OFF-RAMP ↗

After Valentine's run-in with law enforcement, African Americans laid claim to a patch of sand at the end of Pico that came to be known as Inkwell Beach. Along with Bruce's Beach in Manhattan Beach, it was the rare seaside spot in Los Angeles County where Blacks were allowed to congregate. Segregated beaches were invalidated by the courts in 1927, though the practice continued for some years after that.

the nearly seventeen-mile freeway came in at $190 million ($104 million for right-of-way acquisitions, $86 million for construction). From groundbreaking to completion, the whole enterprise took less than ten years, and the bulk of it was built in just four (1962 through 1965). By any measure, it was an engineering success—the rare project that came in under financial and scheduling projections.

As the cherries on top, the Los Angeles City Council honored Marilyn Reece for her "breakthrough" work in "this masculine endeavor," and a prominent cement union bestowed a special "Creative Design" award to the California Division of Highways in honor of her interchange. Unfortunately, the project's visionary female architect was not on hand to accept the bronze plaque—it was presented to three male colleagues—but in 2008 she posthumously received an even bigger honor. The Santa Monica-San Diego node was renamed the Marilyn Jorgenson Reece Memorial Interchange.

DIAMOND LADY IN THE ROUGH

It didn't take long for the Santa Monica Freeway to become the most heavily traveled thoroughfare in the nation—and soon, the world. This appellation was both a blessing and a curse. For one thing, it reaffirmed the need for a west-east artery between Santa Monica and Downtown. On the other hand, it made the freeway ripe for traffic mitigation experiments.

Starting in the winter of 1976, Angelenos began to see notices in newspapers, paid for by Caltrans. Contained in a boxed graphic was a quixotic tease: "OPENING SOON: SANTA MONICA FREEWAY DIAMOND LANE EXPRESS." An illustration showed a stream of vehicles in the fast lane blowing past bumper-to-bumper congestion. Readers were encouraged to send in a coupon to get more information on how they could join a carpool, which back then was defined as three or more people per vehicle.

Carpools first gained popularity during World War II, when the U.S. Office of Civilian Defense urged "car-sharing clubs" to conserve rubber and gasoline. The idea saw a resurgence in the early '70s during the environmental movement. The Federal Clean Air Act of 1970 led to stringent emissions standards to reduce air pollution and protect the ozone layer. (If states didn't comply, federal highway funds would be cut off.) Because cars were the main source of California's infamous smog, the state enacted even tougher automotive laws, one of which was the requirement of catalytic converters. Add to that choking traffic,

oil embargoes, an energy crisis, and driving restrictions caused by gas rationing, and you had a perfect storm stirring by the mid-'70s. Something had to give. That "something," it turns out, was the sacrosanct freedom of the solo freeway driver.

A powerful rideshare poster during World War II, designed by Weimer Pursell. Even the ghost of Adolf Hitler looks like a horrible riding companion.

Caltrans had been contemplating a three-person carpool lane for years. The data seemed unassailable. Every vehicle that carried three individuals would remove two vehicles from the road. By this reasoning, traffic on the 10 Freeway could be cut down by as much as two-thirds. To guard against irregularities in their models, the state would take a slow-and-go approach. The Diamond Lane was to be a one-year pilot. It would run only 12.5 miles—between Downtown and the San Diego Freeway—and cost only $100,000, the price for repainting the fast lanes and putting up some signs. (As a temporary experiment, converting an existing lane was far cheaper than building a new one.) Plus, it would only be enforced during morning and afternoon rush hours. Low-risk, high reward—what could possibly go wrong?

In retrospect, the real question is—what made anyone think it would go right?

Workers laying down the Diamond Lane insignia.

Despite the fact that high-occupancy lanes had never been implemented before on a major highway, Caltrans rammed ahead like a runaway truck through construction cones. The agency ignored local freeway driving habits that had been ingrained for decades. The California Highway Patrol's own studies showed that only 3 percent of cars on the road held three or more occupants. How would officials get two out of every three motorists to give up their wheels when 97 percent of them were going solo? What was the state doing to address a societal system of labor in which millions of people work similar hours at thousands of different jobs scattered over hundreds of communities? For coworkers at the same company, what financial incentives were being offered to encourage pooling, or for their employer to stagger work hours? The most fatal flaw of all was the decision to convert the freeway's left-hand lanes to

Day One of the Diamond Lane—not a good look. An aerial photo captures the wide-open eastbound carpool lane during morning rush hour.

Diamond Lanes, thereby reducing general-traffic lanes from four to three. If you've ever witnessed the outrage after a surface street undergoes a "road diet," you can probably guess where this was going.

The Diamond Lane went into effect on March 15, 1976, at possibly the worst time for any commuter—6:00 a.m. on a Monday morning. Within the hour, the freeway's three regular lanes were swamped with cars while the Diamond Lane remained a virtual desert. Then, in an ill-advised attempt to regulate traffic, Caltrans programmed eighteen-second red lights on some on-ramp meters, backing up streets for several blocks.

Motorists were irate, and insurrection was immediate. One solo driver threw tacks on the Diamond Lane pavement. Another saboteur dumped buckets of beige paint out the window, hoping to blot out the lane's diamond insignia. Single-driver scofflaws weaved into the forbidden lane and then back out when patrolmen appeared. The CHP rang up 268 violators during the 6:00 a.m. to 10:00 a.m. slot alone. Twelve vehicles were wrecked as a result of unsafe lane changes and general rage and confusion, setting a one-day record for traffic accidents.

In the late edition of the *L.A. Times* that evening, the Diamond Lane's first day on the job pushed Patty Hearst's Trial of the Century to the bottom of the front-page fold. Mammoth typeface usually reserved for deadly natural disasters screamed out, "CHA-OS ON A FREEWAY." An aerial picture—miles of gridlock while two lonely cars had the Diamond Lane all to themselves—spoke a thousand words of failure. One commuter summed up the overall sentiment: "This is one of the all-time dumb ideas, anywhere, anytime." County Supervisor Kenneth Hahn declared it a "flop" and called for its immediate repeal. Even a Caltrans spokesman had to admit the freeway had been reduced to "one long parking lot," though he remained hopeful things would get better because "obviously they can't get any worse."

Wanna bet?

In the coming days, traffic remained bad and revolts picked up. A fifty-vehicle caravan called Citizens Against Diamond Lanes barreled along the right shoulders. Drivers hurled beer cans at Caltrans workers repainting Diamond Lane logos. Picketers marched at on-ramps, holding up signs demanding the full return of their tax-paid lanes. Kenneth Hahn joined them, passing out bumper

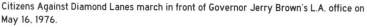

Citizens Against Diamond Lanes march in front of Governor Jerry Brown's L.A. office on May 16, 1976.

Supervisor Kenneth Hahn, foreground, direct-messaging frustrated drivers.

stickers—"Stamp Out Diamond Lanes"—when he wasn't demanding that the city freeze $900,000 worth of bus subsidies (Rapid Transit System buses were also allowed to use the lanes). As a ten-year-old at the time, I vividly remember local TV news running Diamond Lane updates featuring angry protestors, images I had last seen during the Vietnam War. It was my first time hearing the term "socialism"—in this case, a reference to the government forcing people out of their cars and into other people's cars.

Of course, this being L.A., some turned the chaos into art. A popular KFI deejay drove a black hearse up and down the freeway to mourn the death of the fast lane. College kids started charging $1 a pop to sit in a car to satisfy a commuter's three-person minimum. Upon reaching his exit, the driver would drop the students off, and the rent-a-riders would start the process all over again at another on-ramp. Even Johnny Carson, no stranger to L.A. commutes, got in on the fun. As word of the Diamond Lane debacle spread nationally, the *Tonight Show* host worked it into a punchline in a Carnac the Magnificent bit, in which the turbaned mystic would divine the answer before reading a question:

Answer: The Diamond Lane.
Question: What does Zsa Zsa Gabor call the center of a church?

Soon enough, all three broadcast networks were covering the experiment from hell on their national news. Reportage often took on a jocular tone in keeping with the nets' East Coast superiority over alien Western ideals. "Whenever I get back to Los Angeles, I like to check up on the latest contributions to culture and progress," quipped ABC News' Harry Reasoner from a bridge overlooking clogged lanes on the 10. "Past winners include Disneyland, the patty melt, and restaurants shaped like hamburgers. But the latest innovation is meeting with something less than universal approval." Cut to daffy protestors. "Like it or not, diamonds are forever," another reporter shrugged.

But beneath the veneer of frivolity, opposition to the Diamond Lane was serious business. On March 30, 1976, Eric Julber—a Los Angeles attorney—was the first known citizen to file suit against the state in Superior Court. He charged that the lane's three-person minimum discriminated against drivers of two-seat cars. Julber himself owned a sedan, but his complaint was echoed by sports car owners, who suddenly found themselves being punished for dropping a fortune on their Corvettes and Porsches. Less convincingly, Julber asserted that the Diamond Lane violated the Ninth Amendment of the Constitution by forcing him to pick up "strangers" (carpoolers). This denied him his right to privacy. "The whole damn thing is illegal," he concluded.

After several months, enough data had come in to

OFF-RAMP ↗

Most HOV lanes post-Diamond Lane required two people in each car—not three. One exception was the San Bernardino Freeway. One evening in the 1990s, my friend and I were stuck in traffic and slipped into the carpool lane. Finding ourselves one person short, I plucked the hat off my friend's head, stuck it on my fist, and swung my arm into the backseat so that it looked like the silhouette of a third passenger. Other documented ways drivers have tried to fulfill the minimum-passenger rule: dogs, mannequins, blow-up dolls, pregnancy, and, in the case of a hearse, a dead body. The law states a carpool rider must be a person . . . and alive.

reach some conclusions. It wasn't pretty.

Speeds on the Santa Monica Freeway during enforced carpool hours had reduced by 10 percent. Readers writing in to newspapers pointed out that bumper-to-bumper traffic produced more smog and led to poorer gas mileage, which meant that the Diamond Lane failed the very mandate set up by the Federal Clean Air Act in the first place. Injury accidents remained a problem, and not just on the freeway. Thanks to more motorists taking to surface streets, they were up 30 percent on west-east boulevards adjoining the freeway.

Caltrans head Adriana Gianturco.

As spring gave way to summer, tempers reached a new boiling point after the *L.A. Times* published an editorial by Adriana Gianturco, Caltrans's Director of Transportation. Though Gianturco had only joined the agency in March—when the experiment had already been planned for years—she felt compelled to defend it. Her own data showed that it was performing "remarkably well." Bus and carpool usage was inching upward and now accounted for 6.4 percent of all travelers on the Santa Monica Freeway, a figure that was still hard to justify when losing 25 percent of regular traffic lanes. Gianturco went on a full media blitz, proselytizing for the Diamond Lane. But by doing so, she unwittingly walked into a minefield. The Diamond Lane now had a name and face. A scapegoat. A public enemy number one. And, wouldn't you know, she was a woman. Not just any woman—an outspoken one. "My personality is to be direct," she later defended. "To say what's on my mind, not to pussyfoot around."

Ironically, Gianturco had gotten into urban planning because it was one of the few fields that didn't discriminate against women, although, like Marilyn Reece, she encountered mostly men at her job. Her impressive creds included a Fellowship at Harvard and experience with urban renewal projects. Governor Jerry Brown—a classmate from her days at UC Berkeley—tapped her as the head of Transportation specifically for her "new spirit and fresh outlook." As the keeper of a 16,000-mile highway system, Gianturco shepherded the second largest department in the state government. "There had never been a woman in any kind of comparable position in California or any other state before," she reflected.

But now, after fifteen years in public service, Gianturco was reduced to a ball-busting stereotype. Radio hosts, talk-show callers, letter-writers, and columnists substituted her name with epithets—Madwoman of Caltrans, Diamond Lady, Dragon Lady, Public Dictator, Directoress-General in Charge. One conspiracy held that she had named the lane after her diamond jewelry collection (diamonds are often used as symbols for HOV lanes in other states and countries). She was besieged by hate mail and obscene phone calls. Even though she could walk to work, she opted to drive due to death threats. Security started checking her packages for bombs.

Councilman Zev Yaroslavsky at a meeting with his infant daughter in 1978. Decades later, he'd oversee "Carmageddon" on the 405.

Zev Yaroslavsky lobbed a few verbal grenades of his own. The young, savvy city councilman—newly elected to West L.A.'s 5th District—made political hay at her expense, positioning himself as a Hatfield to her McCoy in the whole carpool kerfuffle. It wasn't that he was opposed to ridesharing, it was just the un-American way it was jammed down drivers' throats without due diligence. Gianturco, he scoffed, was "a woman so arrogant that she tried to tell us it was midnight when we could see with our eyes it was high noon." KTTV Channel 11 invited the duo to a televised debate to discuss the merits of the Diamond Lane. They immediately butted heads, each trotting out conflicting statistics. After the taping, they got in a loud argument, during which Zev accused her of "treating people like guinea pigs." The next day's paper showed a photo of them sitting side by side—Zev glowering, Adriana scowling—with a caption that gleefully read: "DIAMOND LANE ANTAGONISTS."

Around this time, Yaroslavsky took Caltrans to court to put an end to this transit tax-waster once and for all. He argued that the state never undertook an Environmental Inspection Review. Caltrans's lawyers countered that the Diamond Lane was "categorically exempt" from the federal Clean Air Act it purported to represent. Judge Matt Byrne didn't buy it. "If the purpose of the project is to affect the environment, how can you bypass the requirements of the environment laws?" he asked. On August 13— the eleventh day of trial—Byrne ordered the program suspended, and a reversion to general traffic for all lanes. Pragmatism had trumped idealism.

Technically, state and federal officials had several more months to mull over an appeal. Zev was aghast by their "bureaucratic arrogance" and threatened to take the case to the U.S. Supreme Court. Governor Brown didn't see the point—protestors were now picketing his L.A. office—and even Gianturco conceded that she probably should've consulted more with local leadership. Prospects for reviving the project officially died in January of 1977.

Unfortunately, vilification of the "Diamond Lady" continued after the court decision. She couldn't win for losing. After a newspaper reporter snidely asked Gianturco if she even owned a driver's license, she responded, "Of course, I have a 1972 Plymouth." Her response got mistyped—the car was called a "1932 Plymouth"—which led readers to believe she was, as she later characterized, "some batty kook who drives an antique car."

By 1981, the state's Senate Finance Subcommittee deleted Gianturco's salary from its transportation budget, and demanded that future Caltrans directors hold engineering degrees. Gianturco saw these moves as direct affronts to her leadership.

Caltrans's credibility also took a beating, their morale along with it. Adriana's own department turned against her. Some prankster hijacked a company computer to display the image of a middle finger thrust in her direction. Though she worked several more years as the transportation director, by the early 1980s, Gianturco had exited public life. She ended up buying and rehabbing historic homes.

As for the Diamond Lane? Southern California's experience with the ill-fated experiment was such a disaster, Caltrans didn't attempt another one for almost two decades in that part of the state. To this day, the Santa Monica Freeway remains devoid of any rideshare or bus lanes. To add one now would require constructing two new lanes, an expensive proposition well over $1 billion.

But in the end, Adriana Gianturco was vindicated. Los Angeles County now has hundreds of miles reserved for HOV, toll, and bus lanes, not to mention light rail. And although you still see the diamond insignia painted on carpool lanes and signs, the term itself has been retired. According to a Caltrans official, "Because Gianturco's impact was so negative . . . we don't want to use the Diamond Lane designation that's still associated with her." Ouch.

But just as old-timers used to say "Xerox" for photocopies, I have found that old habits die hard when referencing carpool lanes in conversation. Or, as Carnac the Magnificent might put it:

OFF-RAMP ↗

Gianturco and Yaroslavsky squared off in a Diamond Lane sequel. In late 1976, Caltrans planned one for the San Diego Freeway by converting the left shoulders to carpool lanes. In some subversive marketing, Adriana's name and a clearly sardonic happy face even graced "coming soon" billboards on the 405. Governor Jerry Brown sample-drove the new lanes with Gianturco, then took a straw poll at a nearby McDonald's. Like Yaroslavsky, people overwhelmingly opposed ridesharing initiatives. Brown nixed the Diamond Lane for the 405 Freeway, and the lanes opened to all traffic.

Answer: He still says "Diamond Lane."
Question: How can you identify an Angeleno born before the 1970s?

A RECORD RESURRECTION

With one catastrophe out of the way, an all-too-real one occurred at 4:31 a.m. on January 17, 1994. Two bridges at La Cienega and Washington Boulevards suffered partial collapses from the Northridge Earthquake. Mike Reynoso and his family were speeding along the 10 Freeway that dark morning when an entire section gave way. Mike kept driving, and the next thing he knew, his sedan was dipping over the severed edge. The only thing that kept the Reynosos from plummeting to a fiery death was the car's metal undercarriage. It got stuck on the ledge.

In a stroke of bad luck, the Santa Monica Freeway was in the midst of being quake-proofed, but Caltrans had not retrofitted this portion yet. The agency quickly accepted contractor C. C.

The collapse of the 10's crossover of Fairfax Avenue.

Myers Inc.'s $14.9 million bid to fix the bridges, to be paid by the federal government. An entire 3.5-mile section was closed, much of it torn down. Normally, a job of this magnitude would take at least a year to complete, but that would not work for a roadway carrying 341,000 vehicles a day. The freeway that was built in record time would need to be rebuilt in record time. Every day it was out of commission would cost L.A.'s economy $1 million in lost production. Consequently, the 10's reconstruction was given priority over all other freeway projects, with C. C. Myers

Buckling on the freeway east of La Cienega, where workers have removed another collapsed bridge.

offered a $200,000 bonus for every day that work came in before the June 24 deadline.

Meanwhile, lookie-loos (this one included) descended on the crippled bridges. Some pocketed rubble, as if claiming their own memento of the Berlin Wall. One enterprising young man, Curtiss Briggs, boxed up chunks of concrete and sold them for $5 a pop (with $2 going to the American Red Cross). During this time, the Diamond Lane made a brief comeback on an eight-mile stretch of the freeway to help with traffic jams. Zev Yaroslavsky, still a councilman, made sure it would go away when the freeway was done.

On the morning of April 12—less than three months after the temblor—Angelenos were pleasantly stunned to find out that work on the Santa Monica Freeway was done. The city celebrated with an old-school dedication. Mayor Richard Riordan, Governor Pete Wilson, and Vice President Al Gore (President Clinton was banking on those California votes for his next term) all jostled for the spotlight. TV crews on the ground and in the sky captured the removal of pylons as motorists burned rubber over

Evening rush hour on the 10, facing west, with nary a Diamond Lane glistening.

the new overpasses, honking and yelling. A Caltrans employee likened it to "the running of the bulls." But perhaps the real winner was C. C. Myers. By working crews 24/7—actual construction took only sixty-six days—they finished the job seventy-four days early. The firm walked away with a $14.5 million bonus, essentially doubling the money it already earned.

To commuters, it was worth every penny. Later that day, they were never more grateful to be, once again, stuck in traffic on the Santa Monica Freeway.

SWITCHING LANES

I started this chapter swooning over the drive through the McClure Tunnel to Santa Monica's sandy shores. But what about *leaving* the beach? Driving east, approaching the Cloverfield Boulevard overpass, a Caltrans sign—still standing as of July 2021—refers to Interstate 10 as the Christopher Columbus Transcontinental Highway, a designation bestowed by the state Legislature during our nation's patriotic 1976 Bicentennial. Motor a couple more miles, however, and you'll come across another sign. Just past the 405, the Santa Monica Freeway becomes the Rosa Parks Freeway (memorialized in February 2002).

Among the sometimes overlapping names of L.A.'s freeways, this one is particularly illustrative. The juxtaposition of Christopher Columbus and Rosa Parks signs on the freeway is symbolically awkward. Columbus enslaved and overpowered a non-white populace—the exact opposite of Parks, who tried to liberate a minority group from white tyranny.

The Rosa Parks section of the freeway covers the nine-mile strip between the San Diego and Harbor Freeways, where the dislocation of Black communities led to the term "south of the 10"—the physical and racial barrier between Los Angeles and what used to be called South Central. Naming this portion after Rosa Parks seems appropriate; it may not reclaim what was lost,

The Columbus sign, on the right shoulder of I-10 while leaving Santa Monica, July 2021.

but it's certainly a reminder of it. The same applies to Christopher Columbus, but for all the wrong reasons.

In our more enlightened era, California has seen the removal of many statues—Junipero Serra, John Sutter, slave-owning presidents, former governor Pete Wilson (who backed Prop 187's targeting of undocumented Latinos), and yes, Christopher Columbus. If you feel, like me, that it's time to "undesignate" the Christopher Columbus Transcontinental Highway, the good news is, so does the County Board of Supervisors. Movements are underway to remove this designation—and the sign—from I-10, if it hasn't already been done by the time you read this.

Whatever your position, it's fitting that the Santa Monica Freeway would be in the thick of such a debate. Freeways trace our history, and the 10 has been at the crossroads of many core issues that divide Americans. Constructed at a feverish pace from Downtown to Santa Monica, it emerged as both a paean to can-do capitalism and a symbol of racial oppression. When its Westside section was expeditiously rebuilt after the '94 earthquake, charges of elitism were levied on the bureaucrats who made it happen. At a time when scores of the less fortunate were desperate for aid to get their lives back together, it was considered almost gauche to immediately throw $30 million at a freeway that would mostly benefit L.A.'s higher-income motorists.

And what lessons did we take away from the Diamond Lane? Objectors insisted it was unconstitutional—many times over. They said it was an attempt by the state (and a dictatorial shrew) at social engineering. It was a violation of our right to privacy, a threat to our freedoms, and, by stripping away a lane that taxes helped pay for, taxation without representation. It was, in short, un-American.

Maybe these crazed protestors were onto something. The Supreme Court famously ruled "One Person, One Vote." Of course, it also later implied that Corporations Are People. Since the majority of Santa Monica Freeway motorists subscribed to One Car, One Person, maybe it's time to declare that Cars Are People, too. I could see the bumper sticker now: One Car, One Vote!

What could be more L.A. than that?

The Marilyn Jorgenson Reece Memorial Interchange, shortly after its unveiling in 1964.

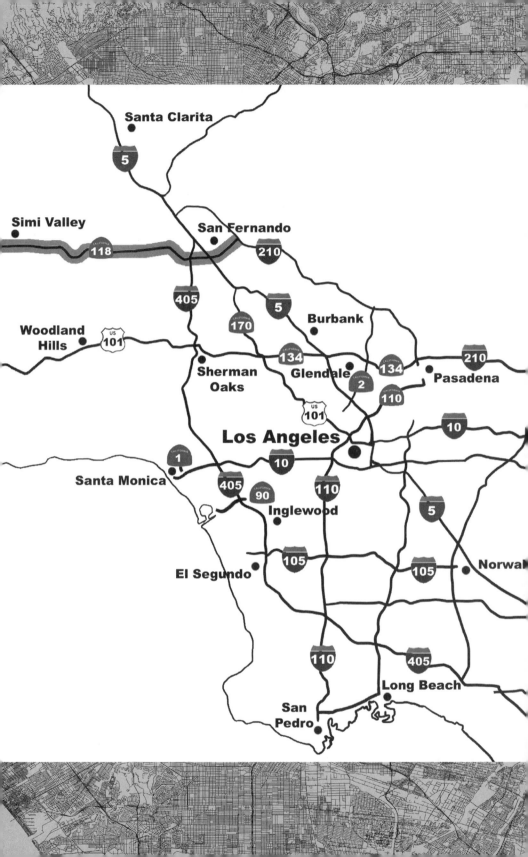

Chapter 10

THE SIMI VALLEY FREEWAY
State Route 118
(1968–1993)

Sadly, a common thread linking the construction of Southern California freeways is that of houses being moved in the name of progress.

But in Simi Valley, houses moved in. Fully formed. Twelve, to be exact.

They were of the pre-fab variety, hauled in by train and wagon in 1888 by a Chicago outfit known as The Colony. Its owners formed part of Simi's first residential neighborhood. Two of the Colony Houses remain standing to this day.

Of course, these newcomers were actually latecomers to the five-by-seven-mile, mountain-ringed valley. For over 10,000 years, the Chumash people had settled in Simi Valley, right up until the early 1800s. As a succession of Spanish, Mexican, and finally American interests took over, Anglos morphed the Indian word "Shimiji"—a reference to wispy clouds—into the more easily pronounceable "Simi." One thing remained constant through its sovereign changeovers: the fertile valley fashioned a

The 118 Freeway, descending Santa Susana Pass into Simi Valley.

reputation as paradise on earth. But then a freeway crashed its pearly white gates. Hello, city slickers from Los Angeles County!

For the first half of the nineteenth century, the Ventura County community of Simi Valley barely registered for most Angelenos. Forty miles northwest of Los Angeles, it was not really on the way to any place. As the only corridor from the upper San Fernando Valley, the giant boulders and craggy precipices of Santa Susana Pass presented foreboding obstacles for motorists. The town didn't even get its first ambulance until 1957, when an auto body repairman tricked out his purple Cadillac with sirens to rush residents to the ER (a veritable superhero, he would slip out of greasy coveralls and into clean whites when his hotline rang). The concept of a Simi Valley Freeway leading to a prestigious presidential library in about thirty years' time would have been unfathomable.

If anything, Simi Valley took pride in its standing as a simple

Despite its unfortunate nickname, "Blood Alley," Santa Susana Pass Road used to be the only reliable passage in and out of Simi Valley from Los Angeles County.

SANTA SUSANA PASS, VENTURA COUNTY, CALIF.

agricultural hamlet. Big dreams came in small but no less meaningful packages: Bumper crops sold to East Coast markets. A new type of fig at the fig festival. High school boys raising hogs and making bacon for class credit. High school girls competing in the Simi Valley Beauty Queen contest ("participants must be between ages sixteen to twenty, cannot be married, and will be judged on figure, beauty, and poise"). The valley was famously suspicious of outsiders, leading to a conservative cultural strain that continues to this day. When state officials proposed a prison in the area, locals argued that housing undesirables would "lower the moral and religious atmosphere of the community." The state looked elsewhere.

Another concern about transplants trampling their Eden was its shortage of water. Lying just over the border of Los Angeles County, Simi Valley was cut off from L.A.'s water grid. As Simi's farming industry grew, once reliable wells went dry, and tank trucks rolled in with fresh supplies of water. In a moment of fleeting desperation, the town even considered annexing with Los Angeles. Officials ultimately decided to import water from L.A.'s Metropolitan Water District, an expensive proposition that cut into farmers' profit margins.

Following predictable patterns after WWII, hordes of returning GIs fanned across the Southland looking to start jobs and families. The San Fernando Valley's housing boom spoked outward, reaching once far-flung communities like Newhall, Thousand Oaks, and Simi Valley. New subdivisions dangled four-bedroom homes with no money down ("Buy now before the Freeway boom!"), and in a time of choking

How happy were happy housewives moving into cheery new homes? Let us count the daisies.

THE SIMI VALLEY FREEWAY

smog, real estate ads promoted "smog-free" living. The massive Rocketdyne complex, which developed and manufactured rocket engines, opened nearby in 1955. Aerospace and defense plants were just "over the hill" in the San Fernando Valley. As property values went up, Simi farmers found that their land would be more valuable if they cashed out to developers. Besides, farming was essentially against the law—water rationing edicts outlawed irrigation. So, one by one, they sold. Over the next two decades, Simi Valley's population would increase from 3,000 to 61,000—a gaudy 1,900 percent.

By 1960, 10,000 commuters meandered through Santa Susana Pass every day. The mountainous two-lane road was notoriously sketchy, earning the sobriquet "Blood Alley." When word got out that a new freeway—the San Diego— would be passing just fourteen miles east of town, Simi residents started singing the

OFF-RAMP ↗

Originally earmarked as the western terminus, Saticoy got shut out of the freeway sweepstakes. As a consolation prize, SR-118—the highway portion of the 118—still starts in that burg, connecting it with Moorpark, where the Simi Valley Freeway begins.

A white line charts the speculative route of the Simi Valley Freeway through upper Chatsworth, March 1965.

freeway gospel. State and county officials were one step ahead of them; in response to population trends, a proposed Simi Valley Freeway began popping up on master plans in the mid-1950s. The fifty-mile route would extend west from the planned Foothill Freeway to Saticoy (near Ventura), a crucial artery that would link the northern San Fernando Valley with the Santa Clara River Valley, much as the Ventura Freeway would connect the San Fernando and Conejo Valleys farther south.

WISH YOU WEREN'T HERE

Not everyone was thrilled about Simi Valley joining the outside world and its lurking undesirables. Older residents feared a freeway would turn it into another San Fernando Valley. Readers carped to the *Ventura County Star-Free Press* about self-interested speculators who didn't understand the pulse of the community. In 1961, a *Footloose*-type brouhaha broke out between the town's religious and progressive factions when a slick developer announced plans to install a cocktail bar in a new bowling alley. It would be the first bar the valley had ever seen, further evidence of societal corrosion. A jam-packed town hall was set up with the Department of Alcohol Beverage Control to debate the divisive issue.

To the Reverend Earl Barnett, the stakes went far beyond bottles of Schlitz. "Anything that weakens our society aids our adversary!" he warned, referring to Russia. He was backed by three Protestant ministers and members of their congregation. They had to feel good about their chances; compared to the Commies, they had God on their side.

ABC rendered its verdict three weeks later: the drunk bowlers won out.

After the California Legislature authorized the freeway in early 1964, it was L.A. County's turn to freak out. Homeowners there became increasingly worried about eminent domain. In

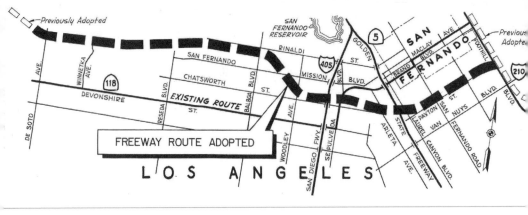

The eastern arm of the planned freeway. The dip in the broken line is to avoid San Fernando Mission, which occupies the space northeast of Sepulveda and San Fernando Mission Boulevards.

the San Fernando Valley community of Mission Hills—which the scheduled freeway path would slice in two—400 families were forced to sell their homes. One mean old cuss refused to let right-of-way agents onto his property. As a result, the feds refused to release a loan to the county until it took control of the holdout's land. But county officials couldn't force a sale without a proper right-of-way survey, which the landowner blocked them from doing. After the circular David vs. Goliath spectacle made the papers, the county made the old-timer an offer he couldn't refuse.

Because eleven miles of freeway would be required to link the Ventura/Los Angeles County line with the 405, transportation officials opted to start on that section first. The logic made sense—upwards of 90 percent of Simi's labor force traveled into L.A. County. On January 4, 1964, 1,500 people—the largest indoor public meeting in the history of the San Fernando Valley—crowded into the Granada Hills High School gymnasium to hear the state's plans for routes through the northern San Fernando Valley. Unlike other freeways, which often mirrored rivers or old rail lines, the path to Simi was not preordained. The Valley was fairly built up by this point and would only keep growing. This meant more voices in the mix. As its northernmost neighborhoods, Granada Hills and Mission Hills wanted to push the freeway as far south as they could. This did not sit well with residents of Northridge and supporters of Cal State Northridge (then called San Fernando Valley State College), which had huge expansion plans to accommodate 30,000 students. Some business owners claimed that the freeway would destroy their livelihoods; others said it would save them. Transportation officials, caught in the middle of a mini-civil war, simply wanted to create the least disturbance to the greatest number of people in the most economical way possible.

As the meeting went on, tempers flared. People shouted over each other. A college official was told to shut his "big fat mouth." Several times, moderator John Bedrosian brought his gavel down with screams of "Order!" But he was shouting into a hurricane. About an hour in, he abruptly canceled the session. The next morning, the *Van Nuys News* led with a full-page headline: "ROW BREAKS UP SIMI F'WY MEET: Jeers, Boos Wreck Discussion of Route."

In the end, engineers settled on a more northerly route, which was partly informed by the state's (doomed) future plans for the Whitnall Freeway across the middle of the Valley. The official groundbreaking of the Simi Valley Freeway occurred during a four-day jubilee kicking off April 23, 1966. Real estate developers pressed flesh, passing out brochures and promising free rodeo tickets for anyone who swung by open houses that weekend. Mountain Valley realtors showcased their new Spacemaker

At a June 30, 1965, San Fernando Valley homeowners meeting, residents overwhelmingly reject the state's bid to build the Whitnall Freeway just a few miles south of the Simi Valley Freeway. The Whitnall would never get completed.

Miss Simi Valley Freeway, center, puts a charge into the freeway's opening with two other beauty queens.

model home, perfect for growing families. Twenty-eight thousand bucks got you a five-bedroom house with a three-car garage. Got another bun in the oven? The Spacemaker had you covered, easily converting from five bedrooms to six thanks to its space-agey use of wall space.

Chatsworth was chosen as the site for the groundbreaking, where the first four miles would roll out through Santa Susana Pass at a cost of $14.4 million. Instead of a customary shovel, dynamite was used to mark the occasion. Sue Welch—aka Miss Simi Valley Freeway—posed in front of a steel-box detonator. Beaming engineers and politicos plugged their ears around her as she set off a satisfying *ka-BOOM!* The press remarked that everyone had a "blast."

Lots of blasts would follow over the next two years. The Santa Susana Pass is among Los Angeles County's rockiest terrain, its sandstone outcroppings a signature feature of the 118

Chumash Indians "in authentic costumes" dazzle the crowd with spears and shields. No peace pipes here.

Freeway. The harsh badlands didn't just draw film and TV studios—*Gunsmoke, The Lone Ranger,* and *How the West Was Won* were among hundreds of productions shot here—but also harbored Charles Manson and his acolytes. The same year that the incendiary cult leader commandeered the dilapidated Western set of Spahn Ranch, explosions could be heard thundering off the canyon walls as the Kirst Construction Company blew up their final rounds of dynamite—7,100 pounds in all. When they were done, 8.5 million cubic yards of earth and rock had been excavated for the freeway at depths reaching 275 feet.

The unopened but completed freeway near Reseda Boulevard provided plenty of pristine concrete for two young rollerskaters in 1982. Watch out for those Botts' Dots, girls!

On July 8, 1968, two years after breaking ground, the four-mile segment opened through the pass. Blood Alley was replaced by a safe, straight six-lane runway. Months later, even more miles were completed, but there was a hitch: eastbound motorists had to exit at Topanga Canyon Boulevard—still eight miles short of I-405. No one could've guessed it would be fifteen more years before the freeway reached the 405 and points east.

MERGING AGENDAS

As the easternmost point, Pacoima represented the 118's eventual terminus in L.A. County. The former farming hub, which gave the world rocker Ritchie Valens, transformed into a convenient suburb for employees of the Lockheed and GM plants in Burbank and Van Nuys. By the early 1970s, Pacoima had become one

OFF-RAMP ↗

On August 25, 2018, the state named the three-mile section of the 5 located south of the 118 the Ritchie Valens Memorial Highway. The Pacoima legend's life was cut short at age seventeen in the 1959 plane crash that also took Buddy Holly.

Ritchie Valens and his band performing "La Bamba."

of the densest neighborhoods in the Valley when right-of-way agents started poking about. Dozens of homes were tagged for demolition. Fears of displacement were augmented by a rise in gang violence, a double-headed scourge that threatened to permanently ravage the close-knit community.

In 1973, Councilman Louis R. Nowell—the same man who went semi-mad by the much-delayed Foothill Freeway—set up a public meeting to debate the freeway's preliminary plans. After witnessing decades of freeway-building in L.A., residents worried that torn-down houses would be replaced by substandard apartment buildings alongside the thunderous road. Besides adding to visual blight, this would lead to health problems stemming from pollution. Studies have found that living within 0.3 miles of a freeway contributes to lower birth rates while leading to higher rates of asthma and cognitive decline in children. Why not build green belts in the form of parks instead? The inconvenient answer was that creating buffer zones did not make economic sense. Caltrans preferred to sell back freeway-adjoining property to private individuals to recoup as much money as possible. End result: Pacoima's needs went unmet while the state parlayed its profits into the next stage of freeway construction or improvements.

If there was a silver lining for residents marked for displacement, it was that Sacramento proposed a two-year freeze on freeway-building in August of 1973. Thank you, Energy Crisis. Gas rationing and OPEC's oil embargo forced the state to rethink all public transportation projects, and the Simi Valley Freeway got swept into the uncertainty. "I think people of Los Angeles are sick and tired of freeways," said Senator James Wedworth. He asserted that state highway officials "don't give a damn about starting a mass rapid transit system. They just want to continue to build more and more freeways that are obsolete before they open and create traffic congestion instead of solving transportation problems."

Wedworth had an ally in, of all people, California Governor Ronald Reagan. Starting in the early '70s, the future president pushed back on proposals to expedite the 118's construction,

Ronald Reagan and wife Nancy in 1972, two years before he would finish out his second and final term as California's Governor.

even vetoing a speed-up bill in 1971. The Legislature found a workaround—a revamped version was signed into law when Reagan was away on business—only to see Governor Jerry Brown kill a similar bill in 1975. Undeterred, the Los Angeles City Council passed a local gas tax hike to raise $200 million for freeway construction projects, including the $90 million needed for the Simi Valley. The freeway even got to piggyback on newly released federal funds. Since the 405 was technically an interstate, engineers were able to tap into that vein to flesh out the Simi Valley-San Diego Freeway Interchange. Brown eventually came around in 1977, signing off on the final five-mile gap of the Simi Valley Freeway at a cost of $35 million. Talk of a bus lane down the middle was scotched.

Despite Brown's late support, transportation officials weren't in the clear just yet. For years, residents, schools, and businesses along the 118 had complained of freeway noise. An environmental impact study determined that the roar of passing vehicles measured an unsafe eighty-five decibels. Legislators forced Caltrans to build sound barriers. Consequently, the Simi Valley Freeway was one of the first freeways in the country to get the precast concrete walls, which knocked noise levels down to seventy decibels. Sound walls are now a regular feature on all Los Angeles-area freeways.

The graceful Arroyo Simi Overhead, looking southwest. The south end of the viaduct marks the transition between the Simi Valley Freeway and the Moorpark Freeway (SR-23).

After all the legislative, political, and environmental maneuverings, the last hole between De Soto Avenue and Balboa Boulevard was not plugged until January 1983. The L.A. portion of the Simi Valley Freeway was now complete, but anti-climactic; the ceremonial dedication had already taken place on December 18, 1982, twice postponed due to El Niño-driven rainstorms. Nonetheless, the 118 could now offer motorists unimpeded access from Pacoima to Simi Valley. West-of-Simi residents would need to wait another eleven years for its hook-up with the Moorpark Freeway (SR-23), two freeways seamlessly blending via the tall, curving Arroyo Simi Overhead—a vital connector to the Conejo Valley and beyond. All told, the Simi Valley Freeway is twenty-eight miles long, its miles divided about equally over two cities—fourteen each in Los Angeles and Simi Valley.

Predictably, traffic volume soon overwhelmed capacity. Ventura County commuters looking to bypass congestion on the 101 took advantage of the 23-to-118 link as an alternate in and out of L.A. County. The opening of Porter Ranch in the northwestern San Fernando Valley also contributed to gridlock. By 1990, vehicles through Santa Susana Pass were exceeding 100,000 trips per day, projected to increase another 50,000 in each of the next ten years. Part of the problem was that the freeway was only six lanes in most areas. Fed-up commuters took to driving on the shoulders. Taking an "if you can't beat 'em, join 'em" approach, transportation officials permitted shoulder-driving as long as it was only during rush hour. Caltrans eventually scrounged up $33 million to widen the roadway.

OFF-RAMP ↗

One unintended consequence of wide-open freeways in the outlying areas of Los Angeles, including the Inland Empire: bank heists. By 1980, the local FBI branch was reporting an average of six bank robberies a day. Southern California's reputation as the Bank Robbery Capital of the World was partly attributable to thieves targeting banks near freeway on-ramps for quick and easy getaways. Added bonus: live-from-the-sky cameos on the five o'clock news.

PARADISE LOST

During a three-year period in the early 1990s, three major happenings visited Simi Valley with repercussions that extended well beyond its tidy borders. The first was the opening of the Ronald Reagan Presidential Library on November 4, 1991. Reagan's preferred location was near his ranch in Santa Barbara, but when that fell through, Simi made a worthy backup. Ventura County was only one county over, and the quasi-Western, Republican landscape aligned with his image.

Securing the bid for a presidential library was quite the coup for Simi Valley. Overnight, the sleepy bedroom community

Air Force One, one of the highlights of the Ronald Reagan Library & Museum in Simi Valley.

bore the mantle of global import. Buses carrying students, tour groups, and senior citizens chugged down the Simi Valley Freeway to check out a slab of the Berlin Wall, ogle a jar of Jelly Bellies, or just pay tribute to the fortieth president. The library also juiced the local economy. Eateries enjoyed a big bump in customers. "We've had people all the way from Orange County and San Bernardino," one restaurant owner gushed. "They wouldn't have been here otherwise in a million years." By April of 1992, 207,000 tourists had walked through the library's doors.

That same month, however, the town's greatest triumph was eclipsed by its darkest hour. On April 29, 1992, a mostly white jury—pulled from different parts of Ventura County—acquitted four Los Angeles police officers of brutality in the beating of Black motorist Rodney King. Though the incident occurred in Los Angeles, Superior Court Judge Bernard Kamins moved the trial to Simi Valley to avoid the media glare. It didn't work. When the verdict was read on live television, much of Los Angeles County erupted in violent outrage, leading to dozens of deaths and the worst urban unrest modern America had ever seen.

Predominantly white Simi Valley—home to more cops than any other similarly sized town in the country—suddenly found itself in the crosshairs. "Burn Simi Valley" was intoned in rap lyrics, on T-shirts, and at protests in front of LAPD headquarters. Simi residents holed up in their homes, afraid to venture out. They were labeled racists. They worried their hometown would unjustly pop up in history books alongside Watts, Birmingham, and other civil rights hotspots. While the city would eventually reclaim its cozy character, Simi lost its innocence that day. The outside world had pierced its bubble and didn't like what it saw. Unfair or not, Simi Valley will forever be bound to Los Angeles and the legacy of Rodney King.

If the King verdict was a figurative seismic event, it took the 1994 Northridge Earthquake to make it literal. Simi's city limits were only six miles from the epicenter. When the 6.7 trembler

Looters set fire to a store on Hollywood Boulevard following the King Beating trial. In addition to the human cost, citywide mayhem resulted in $1 billion in property damage.

struck in the early morning of January 17, Simi Valley sustained considerable damage—$5 million worth to its schools alone. But that paled in comparison to the wreckage sustained by the Simi Valley Freeway where it skirts Northridge. All five eastbound lanes—a 350-foot-long section—pancaked between the Hayvenhurst and Woodley exits. The westbound lanes buckled but didn't collapse. Both sides were torn down and replaced at a cost of $24 million.

As the metro region dug itself out from under the debris amid a crippling recession, it was met by an undertow of nostalgia. The timing seemed right to revisit a popular Californian who declared "Morning in America" during better times. In December 1994, the state Senate voted to rename the Simi Valley Freeway in honor of President Ronald Reagan. It was hardly a slam-dunk. As chairman of the Transportation Committee, Assemblyman Richard Katz (D-Sylmar) held that Reagan's anti-environment gibe from his time as governor—"If you've seen one redwood, you've seen them all"—was disqualifying on its surface. If we're going to designate a freeway, he groused, why not the Moorpark

Freeway, which lies much closer to Reagan's library?

There was, of course, a greater irony at play here. As governor, Reagan had vetoed any efforts to speed up construction of the Simi Valley Freeway. Now they were stamping his name on it. In July of 1995, Caltrans erected ten gleaming green-and-white "Ronald Reagan Freeway" signs along the 118. Its $26,000 bill was paid by the library foundation. Per Nancy Reagan's request, there were no ribbon-cuttings, no beauty queens, no official unveilings. Though he was ill of health by then, Reagan lived another nine years to see his signs before his death in 2004. His body is entombed at the library.

But let's face it. When they aren't calling freeways by their route numbers, Angelenos never refer to them by their "official" names. No one says "hop on the Obama Highway" (a portion of the 134) to Pasadena, or "take the Glenn Anderson Freeway" (i.e., the Century Freeway) to LAX. Renaming freeways after people is a symbolic exercise that, at worst, breeds confusion by working against the pragmatism of freeway signage. It's the reason the Bob Hope Airport reverted back to its original Hollywood Burbank name in 2017. When booking flights, some

Approaching the cut-off for Simi Valley from the 405 North.

passengers had erroneously assumed the airport must be located in Palm Springs, the city most associated with the entertainer (remember the Bob Hope Desert Classic golf tournaments?), whose name also appears on a major street there.

Meanwhile, Simi Valley has come a long way from its early '90s nadir. It remains one of America's safest cities for its size, and polls regularly rank it as one of the "happiest places in the nation" (Disneyland notwithstanding). When driving north on the 405, just seeing "Simi" on Caltrans's green control city signs is strangely comforting. Several things about the word seem to naturally elicit a smile—perhaps the Chumash nod to frilly clouds, the life-affirming "see me" connotation, or something about its phonetics, the way "Simi" and "smile" share three letters.

Now place that word and its positive attributes next to the word "valley," a geographic destination often associated with mythical paradises à la Shangri-La or El Dorado. Like "Hollywood" or "Golden State," "Simi Valley" is more than a destination—it's a dream state. Maybe it really is where the improbable, illusory career of an actor-turned-president was meant to be memorialized all along.

Leaving Los Angeles: The 118 West, entering Ventura County.

Chapter 11

THE MARINA FREEWAY
State Route 90
(1968–1972)

It was mocked as the "freeway to nowhere." It was the subject of a high-profile bribery case presided over by the future judge of *The People's Court*. It changed names more times than Sean Combs. And after finally settling on an official designation, its name was stripped after the humiliation of our nation's thirty-seventh president.

Nevertheless, it persisted.

Though only 2.5 miles long, the Marina Freeway traverses a high-rent district of Los Angeles. Some have likened it to a country club speedway, a gateway to Silicon Beach. But once you dig a little deeper, there's more to this flashy rich kid of a freeway than meets the eye. It has humble origins, literally rising from a swamp—a true rags to riches story. And it's actually quite lonesome, cut off from a long-lost sibling in Orange County. Like twins separated at birth, they belong to the same State Route 90 family, forever banished to the purgatory of separate lives.

Unencumbered by familial obligations, the short-but-sweet Marina Freeway simply exists for your driving enjoyment. Don't believe me? Have a look at its glowing reviews on Yelp (yes, people review freeways on Yelp). Or better yet, simply read on to understand how the freeway no one wanted became the "best kept secret" on the Westside.

The Los Angeles River played an unlikely role in the creation of the Marina Freeway.

The city's famed "mother trench"—which empties into Long Beach—lies seventeen miles due east of the beachside suburb of Playa del Rey. But after severe flooding in 1815, the river changed course and flowed toward present-day Marina del Rey. It maintained this new estuary for ten years before reverting to its old ways, leaving behind the swollen Del Rey and Ballona lagoons, part of a coastal zone of natural springs and *las cienagas* ("the swamps," noted by Spanish explorer Juan Crespi in the 1760s).

As was so often the case pre-freeways, rail provided the first easy access to the wetland. The Pacific Electric Red Car opened up the area in the early 1900s. Before that, in the 1880s, developers partnering with Santa Fe Railroad targeted Ballona as the Southland's main seaport. They had competition: Southern Pacific Railroad and Santa Monica hoped to turn *that* city's bay into a deep-water harbor, going so far as to build the nearly mile-long Long Wharf—the longest pier in the world. In the end, both regions lost out to San Pedro, which in 1897 became the official Port of Los Angeles.

The Ballona watershed may have been impractical as a commercial harbor, but it did provide the perfect backdrop for silent movies. In a Western produced by Thomas Ince, actors dressed

Marina del Rey's former estuary, November 12, 1902. When it wasn't substituting for the Old West for movie directors, it was drawing recreationers and duck hunters.

as Indians paddled canoes through Ballona Creek. In 1915, the legendary filmmaker established his shingle in nearby Culver City to take advantage of the marshland. Soon enough, Metro-Goldwyn-Mayer and other studios settled in the vicinity as well. In 1939, locals recalled the smell of ash in the air, the result of Atlanta being burned to the ground in MGM's *Gone with the Wind*.

OFF-RAMP ↗

Though Howard Hughes's *Spruce Goose* was completed at Terminal Island, the bulk of it was constructed at his Ballona aircraft factory. The billionaire's private airport included the world's longest runway at two miles. Allegedly, Hughes felt that the wetlands provided the perfect surface to slide his huge wooden seaplane into the ocean upon completion. He ended up towing it to Terminal Island instead, and then flew it off Long Beach Harbor.

As with the Los Angeles River, Ballona Creek was prone to flooding, owing to stormwater and perennial streams that emanated from the Hollywood Hills. In the late '30s, it was encased in concrete by the U.S. Army Corps of Engineers. The channel was further shored up in 1959 when engineers raised its levees and dredged silt deposits. Today, the creek is a graffitied concrete wash with a partial bikeway collecting 126 square miles of rain runoff and wastewater from the L.A. basin.

Despite the harnessing of Ballona Creek, the Ballona watershed still occasionally flooded, much to the chagrin of Playa del Rey residents. The main culprit was Centinela Creek. Running just south of Ballona Creek—the two connect near present-day Lincoln and Culver Boulevards—the perennial stream remained wild. While its natural state helped preserve the complex ecosystem of the area, fluctuating water tables and unreliable drainage discouraged industry. As a result, this coastal pocket—west of Culver City, north of Westchester, south of Venice—remained underdeveloped compared to the rest of the booming Westside. That is, until the freeway cavalcade rode in like something out of a Thomas Ince Western.

TELL IT TO THE TV JUDGE

The Marina Freeway was first publicly broached in 1957—before there was even a marina. But Marina del Rey was already in motion thanks to county and federal funding. In September of 1959, the California Highway Commission proposed the adoption of an initial 0.6-mile freeway branching off the emergent San Diego Freeway. Plans were to use the freeway's right-of-way to concrete Centinela Creek and set up a proper storm drain, which would stabilize land values. Playa del Rey residents quickly caught freeway fever. At a public meeting, a citizens' committee urged the freeway's speedy construction, going so far as to say it would accept "any route" that was recommended by the state.

"Not so fast," came a rebuttal from the other side of the room. The row of frowning faces belonged to members of the

This 1947 sketch of Marina del Rey's boat harbor envisioned a more symmetrical orientation offset from Ballona Creek.

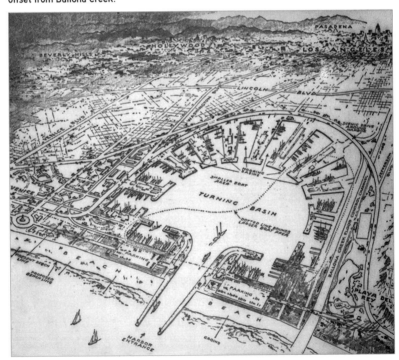

Sepulveda-Slauson Businessmen's Association (SSBA), whose enterprises were clustered around the future footprint of the 405/90 Interchange. "Perhaps the state has a good reason for building this freeway to nowhere," said the SSBA's John Bottle, invoking a phrase that would dog the project for years to come. "But from the information we have at present, we have grave doubts as to its feasibility." The SSBA pointed to the reduction in tax rolls that would result from the estimated fifty businesses that would need to shut their doors. The state countered that the freeway was an investment. The opening of Marina del Rey would mean more businesses and income to the community, with eliminated tax sources offset by plenty of new ones.

The SSBA had an ally in the Culver City Chamber of

An L.A.-vicinity map from August 1962 predicating what freeways would look like in 1980. The Slauson-Marina Freeway is highlighted in gray. Other broken-line routes on the Westside include the La Cienega, Beverly Hills, and Pacific Coast Freeways; in the western Valley, the Reseda and Whitnall Freeways.

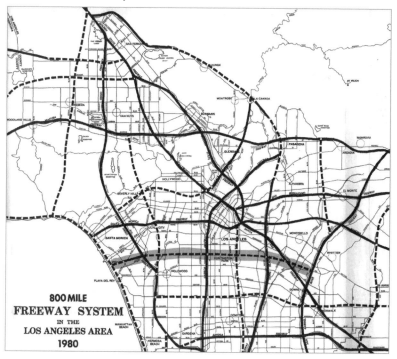

Commerce. It was easy for Playa del Rey residents to champion the freeway from their southerly aeries; they would be far enough away from severe noise, pollution, and displacement. But what about flatlanders, in the finger of Culver City west of the San Diego Freeway? Many would make up the projected 779 families affected by construction. Residents east of the interchange were also concerned. The first 0.6 miles only represented phase one of a planned Slauson-Marina Freeway (or Slauson Freeway, as it was alternately called) whose eastbound passage would blast through the Fox Hills portion of Culver City. And that was just for starters. Ladera Heights, South Los Angeles, Huntington Park, Maywood, Norwalk, La Mirada, Brea—all would soon be carved up like sacrificial lambs for the projected forty-mile, $220 million thoroughfare, which would cross five interstates (the 405, present-day 110, 5, 605, and 710) to its eventual destination in . . . Yorba Linda?

Maybe this really was a freeway to nowhere.

Still, compared to other transportation projects, opposition to the Marina portion of the Slauson-Marina Freeway was almost pro forma. On December 19, 1959—just three months following the first public meeting—the California Highway Commission announced it was moving forward after the County Board of Supervisors and the councils for Los Angeles and Culver City had all waived requests for future hearings. The approved route would start just east of the San Diego Freeway Interchange and proceed west for 3.2 miles, ending at Lincoln Boulevard. Projected costs of construction and property acquisition: $25

OFF-RAMP ↗

Readers of a certain age may recall that Slauson factored into a long-running gag by Johnny Carson. Playing the fast-talking slickster Art Fern, the *Tonight Show* host would offer nonsensible directions to his fictitious shady business, which usually included the line, "You drive to the Slauson cut-off, get out of your car, cut off your Slauson . . ." The studio audience knew this routine so well, they would shout along with Carson: "Cut off your Slauson!"

Ballona Creek's channel and bike path, looking northeast, where it passes under the San Diego Freeway.

million over five years. The Army Corps of Engineers and County Flood Control District were cleared to build the Centinela Creek Channel along the freeway's future path.

Starting both infrastructure projects at the same time was mutually beneficial. Earth from the ditch excavation was used to create freeway berms, and right-of-way acquisitions for both could be done all at once. Thirty-five property owners in the channel's path were given notice to vacate. Ninety-five families were told that their dwellings would be the first to go. As usual, the rights process had its share of hiccups. Mesmer Avenue and Hammack Street became virtual ghost blocks when the state snatched up most of the homes, save for a house owned by a sheriff's deputy named Delmar Epperson. After an extensive survey, the state decided to purchase only the front corner of his lot, which would wipe out most of his driveway ramp. Epperson was told to simply build a new driveway and garage—in his backyard. He was offered $5,500 for his sliver of driveway and attendant property losses. Despite Epperson's protests, the state won out. To this day, the deputy's old house sits on a dead end, wedged against the Marina Freeway. The driveway and garage have been relocated to the back.

Meanwhile, the Sepulveda-Slauson Businessmen's Association still had another arrow in its quiver. Members sponsored an amendment put forth by their representative, state Assemblyman

OFF-RAMP ↗

During excavation for the freeway, UCLA archaeology students discovered bison bones, proof that herds of bison roamed Los Angeles as recently as the fifteenth century. They also unearthed hand axes and pounding stones used by the Tonqva people.

Lester A. McMillan of the 61st District, to stop the "freeway to nowhere" in its tracks. However, it was vetoed by Governor Pat Brown, and no one thought much about this measure or the SSBA again for a few years. Then, in the spring of 1965, the *Los Angeles Herald-Examiner* dropped a bombshell. A Los Angeles County Grand Jury was investigating charges that Assemblyman McMillan had solicited a $10,000 bribe from the SSBA in return for trying to halt the freeway. McMillan called the accusation a "deliberate smear." Adding to the intrigue, the DA's office had in its possession a secret recording from August 13, 1962, that allegedly implicated the representative. Taped by a private investigator, it included a conversation between McMillan and eight businessmen. One of them was Philip Froome, an old friend of McMillan's who ran a shoe store in his district. McMillan was ordered to stand trial in Superior Court for felony bribery, which carried up to fourteen years in prison. In one of those "only in L.A." moments, his judge was the Honorable Joseph A. Wapner. Yes, *that* Judge Wapner, the silver-haired jurist who would later preside over the popular TV show (and Rain Man favorite) *The People's Court*.

The case didn't look good for the defense. McMillan admitted to collecting $10,000 from the businessmen, but gave conflicting reasons. At one point he said he was simply providing legal counsel (the assemblyman was also a lawyer) and the money represented his fee. Another time, he said the cash was a "small salary" for Froome, who had agreed to launch a PR campaign on behalf of SSBA to drum up public opinion against the Slauson-Marina Freeway. McMillan's voice was clearly heard on the tape, telling the men, "I'll give you a run for your dough, I'll promise you that." It's unclear whether McMillan tried to make Froome his fall guy, but by the closing arguments, he had portrayed himself almost as an innocent bystander who merely assisted Froome with political connections. Why a geriatric shoe store merchant with no history in public relations would be calling the shots was another matter.

As with *The People's Court*, this trial was also jury-less. It was up to Judge Wapner to render his verdict. After three days he did: not guilty. Wapner not only bought the defense's argument that McMillan was serving as an attorney for the businessmen, he levied criticism against the state of California. "Few people can live on an assemblyman's salary,"

OFF-RAMP ↗

Judge Wapner appeared in a whopping 2,484 episodes of *The People's Court*, spanning twelve years (1981–1993). Handpicked by Wapner, Rusty the Bailiff was along for every season. Prior to his TV work, Rusty's law enforcement career included guarding Charles Manson in his Downtown jail cell during his circus-like trial.

The Honorable Joseph A. Wapner. Not pictured: Rusty.

he concluded (back then, they were paid about $6,000 plus expenses). "Since the public hasn't decided to do anything about it, it places the members of the Legislature in a terrible position." He suggested that legislators be paid an annual salary that would eliminate conflicts of interest that could arise from taking outside gigs. No word on whether Rusty the Bailiff was present for McMillan's acquittal, although it's entirely possible—Rusty Burrell was a real-life bailiff who served for twenty-five years, including in Wapner's courtroom.

MARINAGATE

Sunset over the marina. And... exhale.

On April 10, 1965, a "ten-day Mardi Gras" kicked off the partial inauguration of Marina del Rey. Events included yacht races, a Naval combat art exhibit, and an appearance by Miss Small Craft Harbor, an awkward allusion to the 780-acre small craft harbor's title as the largest in the world. The $36.2 million development would also include shopping centers, restaurants, and 4,500 apartment units (including some stylin' bachelor pads). Noticeably absent from the fanfare was the Marina Freeway itself. Its first segment was still two and a half years away from opening. But even when completed, it would never quite make it to Marina del Rey—or even Lincoln Boulevard. The SR-90 Freeway technically ends one mile east of the marina, at Culver Boulevard. Consequently, heading west past Culver, the freeway transitions to a forty-five-miles-per-hour expressway that continues 0.7 miles to Lincoln (thus reducing the freeway's true length from 3.2 to 2.5 miles). To rectify bottlenecks at Lincoln, Chairman Burton W. Chace of the Board of Supervisors failed in a bid to span the freeway over Lincoln to Admiralty Way, which would have completely marred the tranquil waterfront. As recently as 2006, the "bridge over Lincoln" idea was resuscitated by L.A. County as part of an elaborate redevelopment of Marina del Rey. (The fact that the county owns the marina was no doubt entirely coincidental.) The idea flamed out after widespread opposition.

The Slauson-Marina Freeway held *its* first dedication on November 22, 1968. The first leg to open was Slauson Avenue to Centinela Avenue—1.5 miles that included the interchange with the San Diego Freeway. In yet another twist in the old-time ritual of ribbon-cuttings, officials hired a helicopter to "snip" a ribbon with its rotors. The freeway's western flank was still a few years off; its Orange County portal, perhaps decades. But one thing was certain—earlier that month, Richard M. Nixon had been elected as President of the United States, and that gave Assemblyman John V. Briggs (R-Fullerton) an idea.

During a legislative session in 1969, Briggs introduced a resolution to name the SR-90 Freeway after Richard Nixon. After all, the freeway's eastern terminus would end in the president's hometown of Yorba Linda. Nixon even *liked* freeways. During his first term, he was known to "cruise freeways" to relax on trips back to his San Clemente manor, enjoying their numbing meditativeness from the cocoon of his Secret Service-driven car, perhaps with a tumbler of Scotch.

Briggs's measure was shot down in the state's Senate Transportation Committee. He took it personally. "The Democrats in the Senate have indicated that California's outstanding native son and President of the United States is unworthy of the tribute our citizens would pay him by naming Route 90 the Nixon Freeway," he fumed. Like Tricky Dick, however, Briggs proved

OFF-RAMP ↗

So why does the Marina Freeway/Expressway abruptly end at Lincoln Boulevard (portions of which are identified at SR-1)? The plan was for a future interchange with the 100-mile Pacific Coast Freeway, which would replace or parallel Lincoln and, more broadly, Highway 1 from San Juan Capistrano to Oxnard.

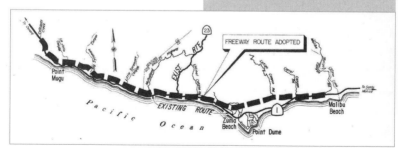

Just a portion of the imagined 100-mile Pacific Coast Freeway, from Malibu past Point Mugu. Oxnard was pegged as the northern terminus. Only a small section of Pacific Coast Highway near Point Hueneme was upgraded to freeway standards (overpasses, access ramps, and embankments).

Future and past presidents—Richard Nixon and Dwight Eisenhower—playing a round in Southern California in 1965. Eisenhower helped create the 41,000-mile national interstate system; Nixon got a 2.5-mile freeway in his honor that was then stripped away.

nothing if resilient. After two more unsuccessful attempts, he finally convinced a more Republican-friendly Senate to approve the name change in April 1971. The former Slauson/Marina/Slauson-Marina Freeway was now officially the Richard M. Nixon Freeway.

The new label did not sit well with many locals. Citizens expressed caution about naming freeways after living people—you never knew how their legacies might end up. Some snarked that the name change was entirely appropriate. "I say let the freeway stand as proof, like the war, that there is no end in sight," wrote *Los Angeles Times* reader Chuck Lerable, whose shot across the bow took down both Vietnam and the freeway. Steve Kommel concurred: "As evidenced by [Nixon's] performance in office . . . this country is truly on a 'freeway to nowhere.'"

Meanwhile, the last piece of the freeway snapped into place on March 30, 1972. The "freeway to nowhere" now at least went *somewhere* without interruption, even if it was only from Fox Hills to Culver Boulevard. Marking the occasion, a platoon of sailboats "sailed" down the freeway on trailer wheels. The *L.A. Times* couldn't resist a dig at the ever-morphing name, inviting readers to call it the Richard M. Nixon Freeway, Route 90, the Marina Freeway, or "whatever name you prefer." (Even the Southern California Auto Club was slow to update changes on its maps.) America, of course, was in the doldrums in the early '70s; the newspaper's stance mirrored the cynical mood of its subscribers, whose warnings proved prescient. By now it was

FREEWAYTOPIA

becoming obvious that the Richard M. Nixon Freeway would never make it to the O.C. More significantly, on August 9, 1974, Nixon resigned in disgrace after the Watergate cover-up.

Mere weeks after Nixon's famous exit from Washington in his presidential helicopter, Assemblyman Charles Warren (D-Los Angeles) proposed changing the route's name to the Gerald R. Ford Freeway in honor of Nixon's successor. An opposing colleague pointed out that the "Ford" name could lead to confusion; there was already a Henry Ford Avenue on Terminal Island. Lawmakers agreed to simply put out a statement congratulating the new president and table the matter for another time.

It would take another year and a half for the California Legislature to officially strip Nixon of his namesake roadway. On March 16, 1976, the *Long Beach Press-Telegram* announced, "NIXON FREEWAY EXISTS NO LONGER." It was to be called, for now and evermore, the Marina Freeway. The change was met with derision among Nixon diehards, who accused the opposition of attempting to rewrite history. Senator James Whetmore (R-Buena Park) called it "another slap in the face." Assemblyman

OFF-RAMP ↗

Nixon fared better in his hometown of Yorba Linda, where the Richard Nixon Presidential Library and Museum now stands. About 2.3 miles of the 90 Freeway was built in northern Orange County—a stretch that still bears his name. But it's never been clear what to call this roadway. Westbound motorists are greeted with a "Richard Nixon Fwy" sign; eastbound motorists get a "Richard Nixon Pkwy" sign. Google Maps refers to parts of this leg as Imperial Highway. So which is it—a freeway, a parkway, or a highway? The mixed signals seem to reflect the contradictions of a complicated man.

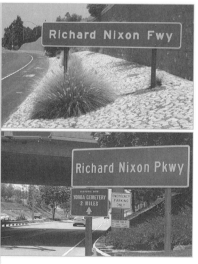

The Richard Nixon Freeway/Parkway/ Highway in Yorba Linda.

John Briggs, who had spent three tireless years campaigning for the Nixon name, also objected. "Nixon has never been convicted of any crime," he groused.

Perhaps both sides could've compromised and just named it the Richard "Not a Crook" Nixon Freeway?

A SUCCESSFUL FAILURE

These days, the tiny freeway that once garnered outsize attention barely registers with most Angelenos—if they even know it exists at all. It has seen one major improvement over the years. As stated earlier, the freeway technically ends at Culver Boulevard. When it first opened, motorists driving at freeway speeds were not prepared to encounter sudden cross-traffic from Culver. After a rash of fatal accidents, County Supervisor Kenneth Hahn pressured the state into spending $3.5 million to build an overpass spanning Culver.

And believe it or not, there was yet one more stab at changing the name. In 2013, based on the recommendation of *L.A. Times* columnist Patt Morrison, Westside Councilman Mike Bonin proposed calling it the Ballona Freeway. "Renaming [the freeway] would allow people to connect the natural beauty around them with the name 'Ballona,'" he said. His proposal coincided with a movement to restore the lagoon's natural habitat. Ballona activists shrugged, with one calling freeways "an artificial thing." At best, they did nothing for the environment. At worst, they destroyed it. Actions mattered more than symbols.

Now, some fifty years after its christening, it's fair to ask: Should the Marina Freeway be deemed a success? After all, it never came close to fulfilling what it was originally meant to be—the start of a nonstop artery from the Westside to northern Orange County. And among the twelve Los Angeles freeways featured in this book, it is by far the least traveled, averaging only about 80,000 users a day.

But those who do drive it love it. And frankly, they don't want you to know about it. Across social networking sites, the word most associated with the Marina Freeway is "secret." But even if the secret's out, I don't imagine much will change. The Marina Freeway is an exception among freeways by virtue of its short length and relative remoteness from central Los Angeles. As a result, it exists in a constant ideal state—one that the founding freeway fathers envisioned for L.A.'s superhighways when they were first dreamed up. One could imagine Senator Randolph Collier and Assemblyman Michael Burns—whose Collier-Burns Act of 1947 created the California freeway system—creeping along the present-day 405, reeling in dismay at its snarling traffic. "What have we *done?*" they'd cry, only to breathe deep sighs of relief after ramping onto the wide-open speedway that is the 90 West. They would experience what many do when chasing a sunset over the marina. Dissipating stress. A feeling of buoyancy. Traffic humming with a free-flowing efficiency that one only fleetingly glimpses on other freeways, usually late at night or during unusual events like a major earthquake, the 1984 Olympics, or a disease that shuts down the world.

The greater Slauson-Marina Freeway never actualized its purpose. But the Marina Freeway, to quote another happy Yelp driver, is a freeway "that actually serves its purpose."

By any other name, that sounds like success to me.

A study in contrasts: Looking west down the 90 (L.A.'s least busy freeway) from the ramp to the 405 (its busiest).

Chapter 12

THE CENTURY FREEWAY
Interstate 105
(1993)

Why do they call it the Century Freeway? Because it took a hundred years to build it.

Hey, I didn't come up with the joke. Variations of it were uttered throughout the thirty-five years it took to finish the eighteen-mile thoroughfare. To put it in perspective, if you added up how long it took to build five other engineering feats—the Golden Gate Bridge, the Panama Canal, the Eiffel Tower, the Statue of Liberty, and Hoover Dam—the number would *still* fall five years short of how long it took to build the Century Freeway.

To be fair, construction comprised only about a third of those three and a half decades. But once it was set in motion in 1958, the actual groundbreaking of the Century Freeway had more head-fakes than Magic Johnson during his Laker days at the nearby Forum.

For multiple reasons, the 105 Freeway is destined to be the last freeway built in the Los Angeles metropolitan area. Retracing its chaotic journey is to chart the rise and fall of the freeway movement. It was rubber-stamped during the freeway craze of the late '50s, then got tripped up by the Environmental Protection Act of 1970. From that point onward, it became an avatar for the soulless freeway lobby, a kind of Death Star against whom the local Rebel Alliance fired back with obstructionism and litigation. And like Han Solo suspended in carbonite, the Century Freeway went into a deep-freeze, only re-emerging in 1982. By that point, it was already the nation's most litigated highway.

When it opened in 1993—all at once, instead of in segments—the old ways of freeways were a distant echo. This was the People's Freeway, the first one designed by lawyers, judges, and sociologists. The fruits of their efforts can be seen in the light rail system down the middle, designated HOV interchange ramps, fewer general-traffic lanes, and housing and job programs to benefit minorities. It was called, if you can believe it, the first freeway with "a heart."

As usual with freeways, aesthetic beauty abounds if you know where to look. Its interchange with the Harbor Freeway—spotlighted in a popular movie musical—is a ten-story stage offering the greatest vistas of any Southland interstate. And while it's true that its C (née Green) Line failed to live up to its promise to deliver passengers right to LAX's doorstep, the freeway itself is a welcome wormhole to the airport that saves drivers from the horrors of the 10 and 405.

If nothing else, let us appreciate the Century Freeway for what it is—the swan song of the L.A. freeway system. Here's to the last dance.

The Century Freeway descends from the finest of pedigrees—Senator Randolph Collier, the "Father of California freeways." On November 7, 1955, Collier and Assemblyman Lee Backstrand made the case for the freeway to the state Legislature's Joint Interim Committee on Public Transportation. From that point onward, the Century Freeway appeared on planning maps in various forms. In one scenario, it would have crept fifty-one miles from the coast to San Bernardino. In another, it would link up with the Yorba Linda Freeway (today's SR-90), before it was decided that *that* route would be assigned to the never-built Slauson-Marina Freeway. By 1958, the state set a course from Sepulveda Boulevard to the Long Beach Freeway (I-710) before extending it to the San Gabriel River Freeway (I-605).

Ironically, compared to other freeways planned at the time, the Century Freeway seemed like it would be the easiest to build. It promised eighteen miles of mostly flat terrain—a straight shot along Century Boulevard (the genesis of its name) to Los Angeles International Airport. Sure, there would be the usual right-of-way issues, but only communists hated freeways in the '50s. They were symbols of free movement, the lifeblood of Los Angeles. As a feeder to jobs in the South Bay, the Century Freeway was predicted to wipe out congestion as the first major west-east artery across the Los Angeles basin. How could it not succeed?

A TRW employee models an Apollo astronaut uniform for school kids in 1969, the year of the first moon landing. The aerospace company owned a 110-acre campus in Redondo Beach known as Space Park.

By the early 1960s, the public was discovering that newly built freeways were not always the answer to perennial gridlock. In fact, they tended to contribute to it. Policymakers urged patience. "People are prone to criticize the freeways because of their peak-hour congestion," observed Bob Brenner, a UCLA traffic expert. "They must remember that the freeways are part of a master scheme that is far from completed. When it is done,

OFF-RAMP ↗

The etymology of Century Boulevard is surprisingly simple. The street is part of a large grid of numbered roads that culminates with 266th Street in Harbor City. Century Boulevard represents where "100th Street" would be.

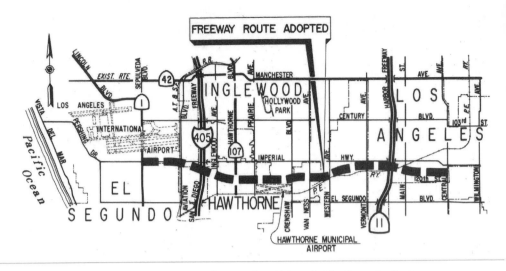

An early schematic for the Century Freeway dipped into central Hawthorne, leading to a contentious battle that would push the 105 north of Imperial Highway.

we will avoid such things as the jam-up at the Downtown interchange and heavy truck traffic on the Hollywood Freeway." Presumably, Mr. Brenner said all this with a straight face.

By 1963, the Division of Highways laid out a principal route. From the east, it would start around present-day Martin Luther King Jr. Boulevard and shadow Century Boulevard. Just past Hawthorne Boulevard, it would channel between Century and Manchester Boulevards and end at the airport. This was the preferred route of Mayor Sam Yorty and the airport commission; to do otherwise, warned LAX General Manager Francis Fox, would make Los Angeles "the only major city in the world not providing direct access

An art teacher and two students visit Watts Towers in 1965, the same year that the freeway commission agreed to create a wider berth around the monument after pushback.

to airport terminals." Fox's prescient comments were dismissed by several parties. Playa del Rey said the route would "kill" its neighborhood, and Westchester residents formed WARR (Westchester Against Red Route, a nod to the color assigned to this proposed pathway). Trans-

OFF-RAMP ↗

In the early 1960s, planners briefly toyed with extending the Century Freeway's western hub from Sepulveda Boulevard to the planned Pacific Coast Freeway, which would have intersected the Century, Marina, and Santa Monica Freeways as it hugged (and desecrated) the coastline.

portation experts worried that a freeway directly under a flight path would distract motorists. When it was revealed the freeway would pass within 150 feet of Watts Towers—the historic monument built by Simon Rodia—the monument's board pushed for "a distance of 700 feet." (The freeway ended up about a mile away.) In a blow to Mayor Yorty, the city council unanimously voted against his preferred route in lieu of a more southerly one. On November 17, 1965, the state highway commission formally adopted this alternate path just south of Imperial Highway, which would be cheaper and displace fewer families. It would cost $150 million to build and, as a state route, would be designated SR-42.

As the state began to acquire properties, an event occurred in 1966 that was a true game-changer. San Francisco often takes the lead on issues of civic activism, and freeways are no exception. On May 21—responding to pressure to preserve the city's quaint cable-car character—its Board of Supervisors voted six to five to reject $250 million in federal financing for two interstates through the city. (This was basically the nail in the coffin for San Francisco's freeway program; seven out of ten were jettisoned in 1959 after a similar revolt.) Within days, California's Public Works Director, John Erreca, flew to Washington with a proposition. Why should freeway-friendly Los Angeles be penalized for its northern neighbor's snootiness? He demanded that the $250 million be transferred to the Century Freeway project. "[Californians'] tax money paid for it and they should get it back," he said. The government eventually approved the entire reallocation to the Century Freeway (the good news was delivered via telegram). Just like that, the freeway became an interstate (I-105) and would be eligible for 92 percent reimbursement from Uncle Sam if completed by 1972. Don't say San Franciscans never did anything for Angelenos.

Now that L.A. had another freeway in the wings, the next step was to sell locals on its eastern passage. Residents from inland municipalities met with state and transportation representatives in packed auditoriums. Battle lines were drawn. Most people endorsed the freeway, as long as it didn't plow through their backyard. South Gate, Lynwood, Bellflower, Santa Fe Springs, and Downey supported the southerly route, while Norwalk and Compton preferred the northerly route.

Other communities saved their wrath for the freeway overlords. Willowbrook—a speck in the crosshairs of the proposed 105/110 Interchange—was located in the "riot zone" of the 1965 Watts insurgency. Residents there felt that the state was purposely trying to "annihilate" their community by gutting the homes of 2,600 families. Hawthorne was also aggrieved. The city had been twice stung—once when the 405 Freeway routed through its western edge, and now with the 105 scheduled to slice through its northern one. Mayor James Q. Wedworth vowed that Hawthorne was "never going to give permits for the freeway." He wrangled a $100,000 war chest to prepare for a legal fight, hoping the threat would impel planners to swing the 105 farther north into Inglewood. But Inglewood didn't want the interstate any more than Hawthorne, and the two cities pushed the boondoggle back and forth between them like a piece of rank fish.

As the state continued devouring properties, it began to receive a rash of bad press for perceived bullying tactics. In the pocket called Del Aire—just west of Hawthorne—residents

The unfinished freeway near LAX, alongside Imperial Highway.

complained of right-of-way agents who pounded on doors while the men of the house were at work. They would wheedle house-wives into unloading their properties well below market value, threatening them with seizure if they didn't comply. Entire blocks were leveled, leaving neighbors "staring, shocked, dazed, and some crying." Esther and Ralph Keith lived in a three-bedroom Del Aire house they had bought in 1947. After an agent buzzed through their home in five minutes and scribbled down his offer, the Keiths refused to sell. Little did the couple know, they would play a crucial role in temporarily torpedoing the Century Freeway.

REVENGE OF THE FREEWAY FOES

In February 1972, the Keiths were among the first to join a class-action lawsuit against California and the federal govern-ment. The plaintiffs were an eclectic group that included three other homeowners, the NAACP, the Sierra Club, and the city of Hawthorne. They called themselves the Freeway Fighters. Their stance was that the Century Freeway was discriminatory, un-justly removing disadvantaged people from their homes without carefully considering environmental or social repercussions. They were represented by four young litigants from the Center for Law in the Public Interest—a nonprofit arm of O'Melveny & Myers, one of L.A.'s largest law firms.

The case went to trial that summer. District Court Judge Harry Pregerson was an avowed apostle of FDR's New Deal who had grown up in Boyle Heights—a tip-off that things might not go the government's way. His forty-five-page opinion was a scathing rebuke of the freeway lobby, whose representatives disregarded environmental impact studies as required by the National Environmental Policy Act of 1970. "The defendants made virtually no attempt to evaluate the effect of the highway on air pollution in the Los Angeles basin," Pregerson admonished. Plaintiff attorney John R. Phillips recited the judge's findings of "wholesale violations" by authorities, including "a failure to make reasonable provision for people who would be displaced by the freeway." Pregerson ordered all work on the freeway to cease immediately.

For Caltrans (as the highway department was now called), the verdict was nothing short of a disaster. The agency had expended $150 million to acquire 4,500 parcels of land. Twenty-five hundred structures had been demolished, and another 660 sat vacant. Twelve thousand people were already moved. The situation was reminiscent of the Long Beach Freeway impasse through South Pasadena—only far worse. The 710 extension never broke ground, but the 105 had already started grading, scarring the landscape of communities along its corridor.

Fortunately, Judge Pregerson was also a pragmatist. He allowed Caltrans to continue buying properties only if a homeowner volunteered to sell. But it would have to meet other conditions to get out of freeway jail. He urged the department to work with the Center for Law in the Public Interest to address housing and economic quandaries created by the freeway and submit a thorough Environmental Impact Report. Only when everything met with his satisfaction would Pregerson lift his injunction.

Of course, 1972 was the same year a judge slapped an EIR order on the Foothill Freeway—a process that froze a five-mile section

Judge Pregerson, in cap, visits a Century Freeway job training site in the 1980s. He was saddened by the abuses of minority-assistance programs he helped set up: "We've got some really deep-seated problems in our society."

of that freeway for almost a decade. But in that situation, crews were allowed to continue building the rest of the 210. Pregerson had locked up the *entire* Century Freeway, with no guarantees that it would ever be resurrected. This created unintended consequences that would only add insult to injury to the corridor cities. Over the years, condemned neighborhoods became ghost towns, haunted by prostitutes and drug addicts. Crime levels shot up. Swimming pools turned into swamps. In Downey, houses were stripped of water heaters, shrubbery, even front lawns, its sod rolled up and carted away. But no municipality was victimized by bureaucratic neglect more than Lynwood.

In 1961, Lynwood was declared an All-America City by the National Civic League. Fueled by the South Bay's aerospace and defense industries, the town was a working-class success story.

Senator Hubert H. Humphrey appraises a mockup of the Space Shuttle at North American Rockwell in Downey, 1972. At the height of the Cold War, the bulk of aerospace companies were based in the South Bay.

Left: A mid-century postcard of Lynwood's main drag exuding post-war promise.
Right: Lynwood in 1980, the bleak cityscape of a "distressed community."

Then everything changed. It was bad enough when right-of-way agents started snatching up homes. Once the freeway stalled, existing homeowners were plunged into perpetual uncertainty. Would the interstate ever be built? Even if they wanted to sell, who would buy a house that might be razed in the future? Meanwhile, Caltrans kept acquiring dwellings, boarding them up like boxy presents to be opened on Christmas morn—or whenever they got the okay to resume work—while arbitrarily leaving others behind.

By 1978, Lynwood had become a hollowed-out hellhole. When they weren't torched by vandals, abandoned homes were overrun by criminals, wild dogs, and, per a city official, "rats big as cats." Fleas invaded intact residences, raising concerns of the plague. One mother dragged her two-year-old son to the Lynwood City Council, lowered his pants, and pointed out all the flea bites on his bottom. Sixty-three-year-old Joseph Spallino, the last occupant on his emptied-out block, slept with a shotgun to ward off intruders, having been robbed seven times since 1975. "I've had nothing but problems since this freeway thing started," he said. "They've taken my heart away. What can I do?" Flagging store sales, a depleted population, and stagnant property values contributed to an estimated $150 million loss in tax revenue since the 1972 shutdown. The Department of Housing and Urban Development delivered the final coup de grâce, labeling the

Evening rush hour traffic jamming El Segundo's streets in 1965. On an average workday, more than 100,000 workers drove in and out of the city, a hub for research, development and manufacturing jobs tied to national defense.

former model municipality a "distressed community."

Business districts were also hamstrung. El Segundo sank hundreds of millions of dollars into a commercial corridor that was predicated on "EZ-FREEWAY ACCESS" to entice tenants. Meanwhile, Hawthorne maintained its longstanding grudge match with the state. City attorney Kenneth Nelson dipped into the war chest to take Caltrans to court and prevent it from condemning any more structures. The agency threatened to simply ram the freeway right through Hawthorne anyway, suspending it on pillars without offering any off- or on-ramps. (Caltrans's

Aerial view of the "Hawthorne Bell," just east of the 105's junction with the 405 Freeway.

Heinz Heckeroth denies this happened.) In 1977, the two parties called a truce. Hawthorne let the freeway pierce its northern border, and Caltrans agreed to bump the roadway northward, over the Imperial Highway, to preserve as many homes as possible. You can identify this section on a map. It's known as the "Hawthorne Bell"—so called because the freeway's curvature resembles the top of a bell.

By the end of 1978, Caltrans attorneys had appeared in court seventy-five times since 1972. Nonetheless, the stalemate between the agency, the government, and Judge Pregerson began to thaw. Caltrans had recently completed an EIR that met with the approval of Carter Administration officials. Extending their federal guideline for funding, Washington reiterated its pledge to tentatively finance 92 percent of the interstate. This was contingent, of course, on an agreement being worked out with Judge Pregerson and the Center for Law in the Public Interest—details that would consume several more years.

On October 11, 1979, Pregerson signed a novel consent decree that laid the groundwork for how the freeway could proceed (the terms would be ironed out by Caltrans and other parties over the next two years). Its scope was unprecedented for a highway project. The state and federal governments agreed to spend $360 million to construct 4,200 lower-income residential units to help replace the 8,100 units removed—the first time interstate funds would be used to build houses. Apprentice and job-training programs would be offered to the 25,000 displaced residents, many of them minorities, who would be allowed to purchase homes in the freeway district at below-market rates. Tutors and scholarships were made available for school-age kids. Transportation officials were ordered to finance brand-new day-care centers (which sounds like the premise for a bad sitcom). The decree even led to housing for AIDS patients in West Hollywood—ten miles away.

Environmental edicts were aimed at minimizing community

disruption. Caltrans agreed to build the freeway mostly above or below grade, with ten fewer on- and off-ramps than originally envisioned. General-traffic lanes would be knocked down from ten to eight to accommodate two HOV lanes. (The Green Line was still a few years off, but light rail was already part of the conversation involving a transitway from Downtown.) Plaintiffs for the Freeway Fighters were ecstatic. The Century Freeway, in the words of lawyer John Phillips, would now be "designed to help rebuild cities, rather than carve up vast sections of the urban area." Judge Pregerson pronounced it the first freeway "that has a heart."

Anyone who has spent time in a boardroom would never associate it with having a heart. Despite good intentions to create what they termed a "humanely sensitive" freeway, Caltrans had its hands full redesigning the 105 and drafting new engineering agreements with the nine corridor cities. The agency was not set up to administer low-income housing and affirmative action initiatives. After the scaling back of funds from the Reagan Administration, Pregerson's decree was tweaked to allow for private-public partnerships and subsidized housing. Court-appointed consultants also recommended a "special master" to help implement Pregerson's directives. But first, freeway officials extended a peace pipe to beleaguered Lynwood, choosing it to host the groundbreaking for the new and improved Century Freeway 2.0.

OFF-RAMP ↗

Like so many other civic heroes in freeway-happy Southern California, Pregerson has been memorialized in concrete. The confluence of the Century and Harbor Freeways is officially the Judge Harry Pregerson Interchange.

The Pregerson Interchange, with the 105 running left to right.

Start 'em up (again): After a decade of court-related inactivity, crews resume work on the Century Freeway in the early 1980s. The delay led to a complete overhaul of Caltrans's freeway-building systems and staggering overruns.

A SYSTEMS SIGALERT

On May 1, 1982, a thousand spectators gathered in a grassy field decked out in red, white, and blue bunting to watch dignitaries dig up the freeway's first spades of dirt. There were free hot dogs, all-girl drill teams, and a patriotic brass band. For one fleeting day, Lynwood was an All-America City again. But beneath the frivolity, Lynwood had a dirty little secret. As Caltrans dug through the city, it unearthed an illegal dump from the 1950s that was oozing toxic waste. The agency spent $80 million and much of the 1980s removing the hazardous heavy metals, trucking them 200 miles away to the what-did-we-ever-do-to-you San Joaquin Valley.

The unexpected outlay couldn't have come at a worse time. Caltrans was bleeding money. Back in 1968, the freeway's cost had doubled to $300 million, a figure that rose to $500 million in 1972. Now the agency was hoping to finish the job for $1.5 billion. Heck, it had already burned through $300 million with nothing to show for it. As gas taxes to fund state highways remained the same, land and construction costs continued to soar. Meanwhile, the court-mandated social programs were, at best, an unknown variable. All this, and Caltrans was also required to issue its final construction contract by September 1990—the new deadline for getting a 92 percent "rebate" from the government's Interstate Highway funds.

FREEWAYTOPIA

By September 23, 1983, Caltrans had assembled an army of seventy-eight contractors to fast-track the Century Freeway. Typically, construction firms were headed by white males. But as part of the federal consent decree, Caltrans was required to assign 35 percent of construction dollars to minority- or female-owned businesses. This was a boon for the diverse owners of smaller outfits—cement pourers, landscapers, sound wall builders, and the like—now suddenly in demand as subcontractors for larger firms. For many, it was a once-in-a-lifetime opportunity to get their foot in the door for larger projects down the line. Unfortunately, even here the reality did not match up to the promise.

Small-business owners were given no instruction manual on how to navigate the federally funded program—a bureaucratic morass of confusing paperwork, bonds and insurance, delayed payments, and often strained partnerships with bigger firms that may have resented hiring them in the first place. When Willa Porter, the Black owner of a modest cement business, was awarded a $50,000 subcontract in the mid-'80s, she likened it to a freedom march: "I thought, 'This is what it must have been like to be a Black in the '60s—if you walk over enough dead bodies, the world changes.'" One year later, instead of making $10,000, she lost $30,000 and put her home on the market. Though she conceded some errors in judgment, she felt she was set up to fail: "You don't get much help from the prime contractor or Caltrans

Minority and women trainees at a Century Freeway job program site.

or anybody else." In a *Los Angeles Times* investigation, a source believed that "at least 80 percent, perhaps as many as 90 percent, of the minority and female businesses that tried to take advantage of the Century Freeway's high affirmative-action goals have gone broke or dropped out of the program."

One path to solvency was to strengthen the relationships between sub and prime contractors, giving the latter a more hands-on approach. But embedding two separate entities carried its own risks. This was all too apparent when Gueno Development Company was accused of being a front for Roadway Construction Company. While there really was a Black-owned company called Gueno, a watchdog group implied that Roadway had paid off its owner, Andrew Gueno, in order to fulfill minority-hiring regulations and get access to a $10 million dirt-hauling job. Roadway even leased office space to Gueno. Caltrans wasn't interested in opening an investigation. "[Minority firms] think everybody with a wheelbarrow and a shovel should have a job," griped a Caltrans representative. "That's not the way it works."

Adding to Caltrans's stress load was the Los Angeles County Transportation Commission. A precursor to the Metropolitan Transportation Authority, the agency took forever deciding what type of transit system to build along the freeway to fulfill the federal mandate. In July of 1984—the same month the Olympic Games launched in Los Angeles—the commission settled on a light rail line to the tune of $254.5 million. While more expensive than a dedicated bus line, rail would be cheaper in yearly operating costs and part of a larger network. Because it would

The center platform of the Willowbrook/ Rosa Parks Station for the freeway's light-rail line in 2014.

Southerly view of the designated HOV lanes connecting the Century and Harbor Freeways. Filmmakers shut down the 105 East-to-110 North flyover ramp in 2015 to shoot the opening musical number of *La La Land*.

be embedded in an interstate, the federal government would kick in its 92 percent. The Century Freeway was now in line to become the first freeway to open with carpool lanes *and* light rail—one that promised to go to LAX. Details on how passengers would get from the end of the freeway to airport terminals were never mentioned, and no one bothered to ask. The city was too busy buzzing on its Olympics sugar high.

Within a few years, two humongous interchanges rose to blot out the sky. The Century-San Diego Freeway Interchange broke ground first, claiming the mantle of the first five-level junction in Los Angeles. The seven-story behemoth had seven miles of connectors and a 200-acre footprint. It was the largest contract awarded by Caltrans—$200 million. "Professionals from all over the world will come to see and study it," said Caltrans spokesman Chuck O'Connell, echoing words from engineers forty years prior, when the first four-level interchange opened Downtown.

But it was the five-level Century-Harbor Freeway Interchange that would draw the most oohs and aahs—and pangs of vertigo—as the tallest nexus in L.A. County (it also contains a rail station platform sandwiched in the middle). The carpool-lane ramp from the 105 West to the 110 North is a veritable thrill ride that offers a jaw-dropping panorama of the Los Angeles basin to the point of distraction. I'm often so taken by this stunning back-drop that I have to remind myself to keep my eyes on the road. Perhaps no one was more relieved to be done with the thing than the contractor, MCM Construction. In the final years before its completion, nine dead bodies had been dumped at their work site.

THE END OF AN ERA

Though the 110 and 405 junctions still had a couple of incomplete ramps, the 17.3-mile Century Freeway officially opened to traffic on October 14, 1993. Several hundred spectators filled bleach-ers at the 105's interchange with the 710. Three Laker Girls, two belly dancers (!), and Kingston, the Los Angeles Kings' snow leop-ard mascot, gyrated to the percussive blasts of the USC Marching Band. Representative Glenn Anderson—who would die a year later due to declining health—raised a plaque that named the freeway in his honor, even though no one before, during, or since

"Ready? Okay!" USC cheerleaders and band members stand by for the 1993 ribbon-cutting on the Century-Long Beach Freeway Interchange. Spectators take in the dedication from the flyover behind them.

OFF-RAMP ↗

California lawmakers passed a resolution naming the Century Freeway after Glenn Anderson in 1987, six years before it opened. As a congressman (D-San Pedro), Anderson helped secure federal funding for the 105. The representative had a personal connection to its pathway; he was born and raised in Hawthorne.

ever referred to it as the Glenn Anderson Freeway. In keeping with the "freeway with a conscience" theme, the dedication included an incense blessing conducted by an American Indian in warrior garb whose face and long hair bore a striking resemblance to Nicolas Cage's in *Con Air*.

Governor Pete Wilson did the ribbon-cutting, scissoring a string of yellow silk poppies before climbing into a vintage auto. (Some rituals never die.) At 4:15 p.m., the barricades were lifted on both ends of the freeway—El Segundo and Norwalk—where eager motorists had been idling for hours.

What they encountered was a freeway like no other: The first to employ three modes of transportation (regular lanes, HOV lanes, and rail). Rideshare lanes that had their own interchange bridges. Signaled meters on connector ramps. TV cameras and fiberoptic sensors embedded in the pavement to monitor traffic. In short, all the things we now take for granted with today's upgraded, modern freeways.

Of course, there were unenviable precedents as well. Thanks to its thirty-five-year journey, the Century Freeway was far and away the most expensive roadway ever built in the United States. Remember its initial $150 million price tag? It ended up costing $2.2 billion, or $127 million per mile (compare that to the $1 million-per-mile of the Arroyo Seco Parkway). The litigation alone resulted in 10,000 inquiries or lawsuits and an Environmental Impact Report that grew to be more than seven feet thick! "There probably won't be another freeway built in an urban area," said Jerry Baxter of Caltrans. "The cost is too high, and the environmental issues are still paramount." From the 1990s onward, transportation officials would focus on mass transit and upgrading existing freeways. As transportation expert Arthur Bauer remarked at the time: "The Century Freeway is a magnificent study of sort of the end of an era and the beginning of another."

For Ester Keith, the old era still left a bitter taste. Decades later, the original Freeway Fighter never forgave Caltrans's "demanding, abrasive attitude" that led her and her husband Ralph to refuse to sell their Del Aire home and force a trial. Ralph died in 1983, and Ester lost her home to Caltrans shortly thereafter, for $12,000 less than its open-market appraisal. Cruelly, her former house was spared after all—part of the Hawthorne Bell that shoved traffic northward—and a developer later swooped in and built a profitable housing complex in its footprint. When asked by a reporter if she would ever drive the Century Freeway, the feisty, chain-smoking septuagenarian replied, "Are you kidding me? The only way I'd go on the Century is if I was being hauled somewhere in my casket." Ester wasn't the only one who planned to boycott anything involving the Century Freeway. When she and another evictee heard about the ribbon-cutting ceremony, he told her, "What I'd like to do is wrap the ribbon around their necks and pull hard."

Beset by its own problems and delays, the Green Line (now C Line) did not open until August 12, 1995. The ceremony featured a trumpeter from nearby Hollywood Park and a soundtrack of Beach Boys music—the latter a sad reminder that the childhood home of the brothers Wilson

---

OFF-RAMP ↗

The 105 famously subbed in for the 10 Freeway in the 1994 movie *Speed*, in which a Santa Monica Big Blue Bus is rigged with a bomb. To keep from triggering the explosive, Sandra Bullock's bus driver must keep her vehicle clocked above fifty miles per hour (the lack of traffic is the flick's most laughable piece of fiction). The filmmakers had six weeks to shoot on the unfinished freeway in exchange for paying some construction costs. They got their money's worth when—with a little CGI assistance—the bus climactically soared over a "gap" in the freeway at the Century-Harbor Interchange.

at 3701 W. 119th Street was torn down to make way for the 105. (The spot is marked by a very cool monument, which includes a frieze replication of the cover art for the single "Surfin' Safari.") Beyond the fanfare, the rail line brought out the critics. Its original $254.5 million price tag had escalated to about $1 billion while projected daily ridership shriveled from 100,000 to 10,000. Officials partially blamed the end of the Cold War for the lower estimates; the shuttering of aerospace and defense contractors resulted in tens of thousands of lost jobs in the South Bay, the Green Line's terminus. Which, of course, begs the question: Why *does* the line hook southward instead of heading into LAX?

The MTA's primary goal was to serve commuters—not airport travelers—although the agency did intend to build a subsequent spur that would extend to the airport's Lot C, where riders could hop on a people mover around the horseshoe of terminals. Officials encountered friction, however, with airport and Federal Aviation Administration officials, who worried that an elevated train would mess with radar systems and distract pilots. Their recommendation was a subway. Running out of money, the MTA dropped the idea altogether.

Make no mistake: The C Line's lack of access to the airport was its original sin. A member of the state Senate called the decision "scandalous." An editorial cartoon in the *L.A. Times* showed a man with two suitcases standing on the banks of a river—where train tracks stopped at the water—staring forlornly at the tantalizingly close LAX Theme Building on the other side. I can't count how many times I explained to puzzled out-of-towners why they couldn't take a train out of the airport. Even MTA board member James Cragin acknowledged, "It's kind of like having an elephant with no tail and no trunk." Indeed, the incomplete elephant in the room even tainted the touchy-feely Century Freeway. Both the train and the freeway are often mocked as routes from "nowhere to nowhere" (which also alludes to the fact that neither connects to the 5, where there is also a major rail hub). But their

The Automated People Mover (APM) will ferry passengers to six stations in and around LAX.

reputations are on an upswing.

Thanks to L.A. County voter-approved sales tax increases, the MTA finally has the resources to provide rail connections and people movers to LAX. There are also plans to extend the C Line farther east, where it would finally connect to Metrolink and Amtrak. The Century Freeway itself has proven to be way more than a road to nowhere, handling over a quarter-million daily vehicles for most of its run while providing a vital link to the airport from Downtown and points east. It also stands as a testament to the power of community activism. Led by the Freeway Fighters—Judge Pregerson, the Center for Law in the Public Interest, and homeowners like the Keiths—the Century Freeway represented a paradigm shift that began with the Foothill and Glendale Freeways. Gradually, the dictatorial days of highway officials determining routes was replaced by consensus building, resulting in more socially responsible public-works projects.

The Century Freeway is not so much the end-of-the-road of old practices, but rather a closing of the loop. Half a century before the Century Freeway, the citizens of Los Angeles and Pasadena sped down a newly opened high-speed, functional parkway that was also environmentally sensitive and aesthetically pleasing. But it would prove an anomaly. Starting with the Hollywood Freeway, thoroughfares prized speed and efficiency. Interstate

105 had been on this same track in the 1960s, until federal environmental and housing laws caught up with it. After its ten-year timeout, the freeway's reboot in 1982 was guided by the human element. Quality of life, including the incorporation of more green space, became a central issue of future civic infrastructure projects—a philosophy codified in this quote: "Without adequate parks, the bulk of the people are progressively cut off from many kinds of creation of the utmost importance to their health, happiness, and moral welfare."

The passage is from 1930, when Harland Bartholomew and the Olmsted brothers issued their master plan to the Los Angeles city leaders. Their vision greatly informed the design of the Arroyo Seco Parkway. Like its early predecessor, the Century Freeway was built—or shall we say rebuilt—from scratch to be something more than a freeway. While the journey was far from perfect, the progressive social reform of L.A.'s first freeway could be seen in the guiding principles behind L.A.'s last freeway.

The circle was now complete.

The beginning is the end: The western terminus of the Century Freeway, throwing shade over Imperial Highway.

END FREEWAY

As a child on family road trips, I used to lean forward, seatbelt casually unbuckled by my side, in anticipation of the end of the freeway. Interstates seemed to stretch on with an infinity akin to the universe.

But I rarely got to witness the end of the space-time continuum. Our trusty Ford Gran Torino Station Wagon usually rocketed off a random exit for some other destination, leaving me slumped in disappointment. On those rare occasions that I glimpsed the end of a freeway through our car windows, there was something about those "END FREEWAY" signs—with their screaming all-caps black letters and blinking yellow lights—that gave me the same fleeting rush as glimpsing a streaking meteor against the night sky or a black bear lumbering past our car in the Sequoias.

After decades of dominating the Los Angeles basin, we may be seeing the end of freeways, certainly in the metaphorical sense. Even Caltrans veteran Heinz Heckeroth—ninety-five years old as I write this—admits that the state agency is no longer in the freeway-building business. "Maintain and operate," he told me somewhat wistfully. "That's all it is now. It's completely changed."

In the short-term future, Angelenos can expect more HOV and HOT (High-Occupancy Toll) lanes, and other traffic measures rooted in congestion pricing. Longer-term visions involve erasing L.A. County freeways from the landscape altogether—or at least segments of them. I already covered the proposals to convert the little-used Terminal Island Freeway and the Glendale Freeway stub near Silver Lake into elevated parks à la New York's High Line. A grassroots campaign to reimagine the condemned Riverside-Figueroa Bridge as a pedestrian zone also got some noise before failing. "Cap parks" over trenched freeways remain part of the conversation. A prototype already exists where Memorial Park covers the 210 in La Cañada Flintridge. Younger Angelenos are eschewing cars and driver's licenses in favor of better public transit and safer roads for biking. And if the COVID-19 pandemic taught us anything, it's that many businesses are now equipped to accommodate the work-at-home model for white-collar jobs. Why commute at all when you can log in from a cabin in Joshua Tree?

The trend away from freeways really comes down to a yearning by Angelenos for cleaner air, more open space, and reconnecting with our natural landscape. Not just reconnecting with it, but reclaiming it. In November of 2020, twenty families under the guidance of Reclaim and Rebuild Our Community illegally moved into empty homes in the El Sereno district near Downtown. The residences were among 200 in the area owned by Caltrans, expropriated for the never-built 710 Freeway extension.

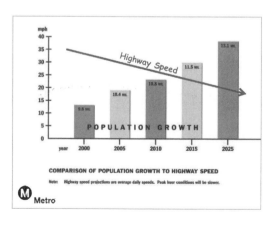

COMPARISON OF POPULATION GROWTH TO HIGHWAY SPEED

Note: Highway speed projections are average daily speeds. Peak hour conditions will be slower.

Metro

Though the agency allowed other tenants who seized Caltrans homes earlier in the year to stay and eventually pay

Per the MTA, average daily speeds on L.A. County freeways continue to trend downward as the population grows.

Following George Floyd's death while in police custody on May 25, 2020, hundreds of protestors took to the 101 Freeway at the Downtown Slot and here, on the southbound lanes through Hollywood.

rent, it deemed the current structures "unsafe and uninhabitable." CHP officers forcibly removed the trespassers during Thanksgiving Week to widespread criticism. To Pastor Michael "Cue" Jn-Marie, who helped organize the movement, the vacant state-owned residences are residues of the disruption that freeway planners inflicted on a tapestry of communities—which, frustrated by lack of progress, have taken the initiative to stitch themselves back together.

As for the freeways that *were* built, 2020 also saw Black Lives Matter demonstrators occupying the Hollywood Freeway in the aftermath of George Floyd's death. Pastor Cue believes the public should brace for more. Los Angeles freeways are ready-made symbols of systemic racism—or, as he puts it, "the arteries that carry the blood of segregation through America." Any inconveniences to commuters are the point. "We're sending a message to the suburbs," Jn-Marie says. "You're affected by the freeways, but we were affected by the freeways a lot longer than you were. It created an inconvenience for us in just being able to survive."

As previously observed, freeways are living monuments reflecting the evolution of Los Angeles. The operative word here is "evolution." The L.A. metropolis is ever-evolving, reexamining the behaviors and policies of yesteryear, course-correcting for

FREEWAYTOPIA

the future. To their credit, the California Department of Transportation is not hiding from its history. In a statement put out on December 10, 2020, Caltrans acknowledged that "communities of color and underserved communities experienced fewer benefits and a greater share of negative impacts" from freeways, which "quite literally put up barriers, divided communities, and amplified racial inequities." Going forward, it's clear that state and municipal policies will be informed by this reckoning with the past.

Speaking of change, can you imagine a beauty queen—teetering in high heels with giant scissors as she's ogled by a wall of older men in business suits—christening any kind of freeway opening these days? Even the century-old Miss America competition did away with the swimsuit portion in 2018. Against this broader backdrop, debates now openly rage that pit art against artist. Can we still enjoy "Ride of the Valkyries" by virulent anti-Semite Richard Wagner? What about Michael Jackson's music? *Gone with the Wind*? A monologue by Louis C.K.?

I guess what I'm posing is this: knowing their role in marginalizing generations of Angelenos, can we still cherish the aestheticism of freeways, their harmonious interplay of form and function? I like to think our brains are complex enough to still appreciate the grandeur of freeways while acknowledging

Demonstrators against the Vietnam War, taking over the 10 Freeway East near Cal State Los Angeles in 1970.

the troubling context of their origins. I can't help but thrill at the graceful lines of a Marilyn Jorgenson Reece Interchange, the ornateness of the Arroyo Seco Parkway, the unfurling tableau of the Valley as you descend the 405, and yes, even the blinking lights of an "END FREEWAY" sign—every piece fitting squarely into the complicated narrative of Los Angeles like a 3D jigsaw puzzle.

Like it or not, freeways are here to stay. Perhaps it's our relationship with them that could stand changing. If you have a car, embrace the freeway as a bridge to your neighbors. Hop on one that might be out of your way. See L.A.'s urban sprawl through a different lens instead of just blurring past it. Get off and get lost. Explore a community in the freeway's roaring shadows. Patronize its shops, parks, and restaurants. *Feel* the city. Engage. Take a different freeway route home. Suddenly, your world seems a little bigger. More connected. Lower your windows, let in the warm air ... wait—what's this? What's with all the brake lights? The entire right three lanes are blocked up ahead. It's a SigAlert.

Looks like you're going to be here awhile.

The San Diego Freeway (what else?).

NOTES ON SOURCES

California Highways and Public Works informed much of my research for this book. Beginning in January 1924, the bimonthly periodical was the official publication of the California Highway Commission, each issue a boon of information about the state's roadway projects. Largely geared for Division of Highways employees, it was also a handy resource for media outlets to relay freeway updates to the public. The entire catalog has been archived in Caltrans's Transportation Library, functioning as an indispensable repository for freeway stats, budgets, maps, and milestones.

The final *California Highways* edition was January-February 1967. After a sixteen-year hiatus, the department (then called Caltrans) published an employee newsletter called *Going Places*, which ran from 1983 to 1993. These also proved useful and can be accessed via the agency's library. Back issues of *Public Roads* magazine—put out by the Federal Highway Administration—were also helpful, though their articles skew more toward interstates. Its digital catalog covers 1993 to the present.

Caltrans's Transportation Library also contains oral histories. Primarily interviews with past engineers and designers, they help provide context to California's freeway system.

Other invaluable information across all chapters comes from CAHighways.org, SoCalRegion.com, and Interstate-Guide.com. The Los Angeles Almanac Online and the Census are my primary sources for population trends. Traffic numbers from prior decades are cited from a variety of media and tend to be fluid; occasionally, my figures represent an amalgam from reputable newspapers. Modern-day traffic is easier to accurately track thanks to data from dot.ca.gov and INRIX.com.

BEGIN FREEWAY

Freeways for the Region, Regional Planning Commission, County of Los Angeles, 1943, infuses much of the introduction. Ditto *Caltrans: Then and Now 1895–1995*, a film put together by Caltrans's Steve DeVorkin based on a slide show by Structural Engineer Norman Root.

Richard C. Miller's superlatives about freeways are oft-quoted, including in his obituary on RichardMiller.com. Reyner Banham's *Los Angeles: The Architecture of Four Ecologies* was published in 1971 by the University of California Press.

The SigAlert sidebar is largely sourced from "SigAlert May Stand Ready to Serve U.S.," by Sue Reilly, *Los Angeles Times*, April 25, 1965. (Fun fact: In 1997, Loyd C. Sigmon, well into his eighties, could be seen tooling around L.A. in a Lincoln Continental sporting a SIGALRT vanity plate.) "How Los Angeles Traffic Snarls Spawned a Word," by Todd S. Purdum, *New York Times*, May 18, 1997.)

The sidebar about inserting "the" in front of freeways is informed by my interviews with Heinz Heckeroth, the former Chief Deputy Director of Caltrans, as well as KCET.org: "The 5, the 101, the 405: Why Southern Californians Love Saying 'the' Before Freeway Numbers," by Nathan Masters, November 10, 2015.

My examination of Germany's autobahn system is pulled from various sources, including the *St. Louis Post-Dispatch* (August 15, 1936) and the *Manchester Guardian* (March 12, 1936). The Federal Highway Administration covers Hitler, Eisenhower, and autobahns in an online article called "The Reichsautobahnen," which is an excerpt of "The Man Who Changed America, Part I," by Richard F. Weingroff, from *Public Roads*, March/April 2003.

Robert Moses's parkways have been scrutinized by several transportation experts. Heckeroth spoke to me at length about them. Futuristic speedways, complete with cloverleafs, were promoted by General Motors Corporation in its short film *To New Horizons*, which is viewable online. It premiered at the company's "Futurama" exhibit at the 1939-1940 New York World's Fair.

Interregional, Regional, Metropolitan Parkways in the Los Angeles Metropolitan Area was drafted by the Los Angeles Metropolitan Parkway Engineering Committee, March 30, 1946.

The Collier-Burns Act is easily accessible as a signature of California's legislative history. The Federal Aid Highway Act is viewable on the website for the United States Senate. Heckeroth weighed in on both the rise of California's freeways and federal interstates. Quotations from Eisenhower are pulled from mid-1950s publications, including, for the sidebar, "President Asks For Road Net As Must In Atom Evacuations," by Herb Altschull, *Sacramento Bee*, February 22, 1955.

John F. Kennedy's "Victory has a hundred fathers" line was uttered by the president himself, his audio included on NPR's *All Things Considered* about Bob Shrum's book, *Serial Campaigner* (2007: Simon & Schuster).

The freeway musings of Kurt von Meier are available on the Larry Barnett-curated KurtvonMeier.com.

The Southern California Automobile Club information here and throughout these chapters is based on online archives and conversations I had with Morgan Yates, the Auto Club's resident archivist.

Several sources point to Cobb salads, gangsta rap, and the modern skateboard as having been invented in L.A., among them (respectively): "Cobb Salads," *Los Angeles Times*, July 30, 1998; the gangsta rap entry in *Britannica*;

and the 2001 documentary *Dogtown and Z-Boys*, directed by Stacy Peralta, as well as numerous skater magazines.

CHAPTER 1: THE ARROYO SECO PARKWAY

Much of the Arroyo Seco Parkway's history is derived from "Arroyo Seco Parkway," *The Historic American Engineering Record*, HAER No. CA-265. Also illuminating this chapter are "Freeway dream that became a nightmare," by Al Martinez, *Sunday Star-Bulletin*, March 25, 1973, and "Fifty Years of Moving History," by Patt Morrison, *Los Angeles Times*, December 21, 1990.

The term "stopless motorway" appears in several periodicals from the 1920s, including the *Whittier News*, "To Speed Up Touring," September 27, 1924.

The Los Angeles Conservancy and BridgeHunter.com provide information on the parkway's bridges.

Katharine Hepburn's four Oscars are recorded by the Academy of Motion Picture Arts and Sciences; the four landmark statuses for the freeway are public record, confirmed by, among others, the Arroyo Seco Foundation.

The Smithsonian Institute provides a comprehensive history of the modern-day bicycle and the bicycle craze of the 1890s. Much of my research for the California Cycleway relies on the Pasadena Museum of History. Interestingly, while I can't vouch for its veracity, a reader named Dennis Crowley wrote a letter to the *Los Angeles Times* on July 31, 1999, in which he claimed that Henry E. Huntington "blocked" Dobbins's venture when he "realized the cycleway would be faster than his Red Car trolleys and endanger his growing transportation monopoly."

The opening of the Wuppertal suspended railway is based on various essays, including "March 31, 1901: Wuppertal Monorail," by Keith Barry for *WIRED*, March 31, 2010. John Fawkes's misbegotten monorail from 1907 is recapped on the *Smithsonian* magazine's website.

The *Pasadena Evening Post* posted updates on the Pasadena Rapid Transit Company throughout 1919.

Per-capita automobile ownership, the growth of L.A.'s motorists, and other data are sourced from "Arroyo Seco Parkway," *The Historic American Engineering Record*. The *Los Angeles Times* mentions Pasadenians owning the most cars in the world in "Pasadena bridge spans 100 years," by Joe Piasecki, June 22, 2013.

City planning notations for 1924 are from *Major Traffic Street Plan*, by Frederick Law Olmsted Jr., Harland Bartholomew, and Charles H. Cheney, 1924.

Descriptive accounts of the Arroyo Seco Parkway's construction are gathered from various newspapers, including "High-Speed Boulevard Projected," *Los Angeles Times*, August 20, 1934 (the source of quotes).

The Los Angeles Conservancy is my primary resource for the City Beautiful movement.

Early *California Outlook* issues explicate efforts to create lakeside byways described in the sidebar.

The interchangeability of "parkway" and "freeway" nomenclature was first evident in the early 1930s, including in a United Press wire story titled "Freeway, New Type of Automobile Road, Proposed" for the *Ventura County Star*, April 24, 1933. The technical definition of a freeway is articulated on many states' department of transportation websites, including dot.ca.gov.

Details about the state's first gas tax are derived from the Legislative Analyst's Office (LAO). For federal financing of the parkway, I consulted several sources, including KCET.org, "Departures: Highland Park: The History of the Arroyo Seco Parkway," October 24, 2011.

In addition to *California Highways and Public Works* archives, my portrait of Lyall Pardee is drawn from "Pardee Made Permanent City Engineer," *Valley News*, December 27, 1956, and "Ex-City Engineer Lyall Pardee Dies," by Michael Liedtke, *Los Angeles Times*, December 29, 1982.

Right-of-way information through Elysian Park was analyzed by the *Pasadena Post* and *Los Angeles Times* in the 1930s.

Pastor Stephen "Cue" Jn-Marie spoke with me about race and freeways.

Frank Balfour figures prominently in *California Highways* magazines. I also source "Balfour Retires," *Pomona Progress-Bulletin*, December 23, 1960.

For the sidebar, the March 25, 1937, edition of the *Pasadena Post* mentions the "Freeway Truth" leaflets, their cause further outlined in "No 'Stooges' Employed Save To Shift Auditorium Scenes," *Pasadena Post*, March 26, 1937. Opposition to the freeway extension was rampant per various accounts, including "Boulevard Plan Rapped—Proposed South Pasadena Freeway Routing Draws City Manager's Fire," *Los Angeles Times,* December 8, 1935. "Pasadena Freeway Future Uncertain," *Pasadena Independent*, January 2, 1960, speaks of the freeway not meeting federal standards.

Justus Craemer's test-drive appears in "'Along El Camino Real,' with Ed Ainsworth," *Los Angeles Times*, March 29, 1936.

In addition to *California Highways* records, Patt Morrison brings to life the "soft" opening for the *Los Angeles Times* ("50 Years of Moving History") on December 21, 1990. Her story also outlines quirkier elements of the early parkway, including the proposal for coin-operated gas pumps (sidebar).

The grand opening, and Governor Olson's words, are drawn from several local papers including the *Los Angeles Times*, "Freeway Dream That Became a Nightmare" by Al Martinez, March 25, 1973. In a cute aside, when Rose Queen Sally Stanton became a mother, her two young children used to call the Arroyo Seco Parkway "mama's freeway." "50 Years of Moving History," by Patt Morrison, *Los Angeles Times*, December 21, 1990.

The Great Dictator premiere is sourced from the *Los Angeles Theatres* blog and "Chaplin at Premiere of 'Great Dictator,'" *The Gazette (Montreal)*, October 17, 1940.

Chief Tahachwee's blessing of the freeway: "Indian path route of freeway 30 years ago," *Escondido Daily Times-Advocate*, April 14, 1968. His "transference" of rights: "Pasadena Freeway Grows," by David Jensen, *San Bernardino County Sun*, April 23, 1968.

Information about the "Kawei" tribe: "Pasadena and Glendale Forgotten Residents," by Sandra Moore, ColoradoBoulevard.net, October 7, 2015.

C. E. McBride's impressions of the parkway ("Sporting Comment") are printed in the January 8, 1941, edition of the *Kansas City Star*.

The intersection of Figueroa and Sunset Boulevard—just farther south of Figueroa and Riverside—was apparently the busiest in the world in 1948, according to C. W. Cook, a surveyor for the Hollywood Freeway. "Route Decision About Hollywood Freeway Waits," *Valley Times*, March 9, 1948.

There are numerous essays on the Figueroa tunnels, including "Before the 110 freeway, Figueroa Street Ran Through These Tunnels," by Nathan Masters, KCET.org, July 14, 2014.

The Historic American Engineering Record, HAER No. CA-265 charts the different "feel" of the southern portion of the parkway.

The full capacity of the Four Level was announced in "Giant Freeway Interchange Ready Tuesday," *Pomona Progress-Bulletin*, September 21, 1953. The name change a year later: "Pasadena Freeway Name Approved," Associated Press, *Daily News Post*, August 18, 1954.

In addition to *California Highways*, upgrades to the freeway are sourced from "Safety Program to Cut Accidents on Arroyo Seco Freeway Outlined," *Los Angeles Times*, March 22, 1948, and "State Considers Action on Hazardous Ramps," by Don Snyder, *Los Angeles Times*, May 11, 1969.

Twentieth Anniversary accounts: "Freeway's 20th Year Observance Slated," *Pomona Progress-Bulletin*, December 25, 1960; "Pasadena Freeway After 20 Years," *Los Angeles Times*, December 31, 1960; "First Freeway Hailed on 20th Anniversary," *Los Angeles Times*, December 30, 1960. Sally Stanton's life: "Sally Stanton Rubsamen, 1941 Rose Queen, dies at 92," by Claudia Palma, *Pasadena Star-News*, April 22, 2016.

Harrison R. Baker's bullish assessment is portrayed in "Highway Safety, Beauty Stressed By Baker, Advocate of Freeways," by Dave Swaim, *Pasadena Independent Star-News*, September 18, 1966. Baker was such a strong advocate of freeways, he staunchly rejected widely held views that freeways "cut towns in two or add to smog," as detailed in "Pasadena Urged to Back Freeways; Harrison Baker, Pioneer of System, in Plea to Leaders,"

Pasadena-Independent, July 14, 1965.

Richard Ankrom's caper was blanketed by news sources. Two noteworthy ones are "How One Fed-Up Dude Fixed an Awful Freeway Sign Himself," based on an interview with Ankrom and Amy Inouye by David Weinberg for *Gizmodo*, February 17, 2015, and "The Fake Freeway Sign That Became a Real Public Service," by Allison Walker, *Good*, January 11, 2010. Ankrom's own account and photos are available at Ankrom.org.

In the sidebar profile, Dennis Crowley is featured in "Pedaling His Bikeway Plan," by Susan Moffat, *Los Angeles Times*, April 29, 1995.

Nostalgia about the 2010s was my own observation, as well as that of Olivia Ovenden in "The Pull Of The Past: How We All Got Hooked On Nostalgia In The 2010s," *Esquire* online, December 24, 2019. The rebrand of the freeway is drawn primarily from *Arroyo Seco Parkway Corridor Management Plan: A Rehabilitation and Preservation Plan for Southern California's Most Historic Road* (Working Draft), February 12, 2004, which was prepared by the National Trust for Historic Preservation, Rural Heritage/Historic Roads Program, and other agencies.

The disparagement of the Pasadena Freeway was gleaned by me from anecdotal experiences and spelled out in "Freeway has seen better days," by Francisco Vara-Orta and Andrew Blankstein, *Los Angeles Times*, July 18, 2008. Those who clamored for Caltrans to preserve the integrity of the parkway are profiled in "Preservationists: Don't change oldest freeway in the West," by Daisy Nguyen, Associated Press, *Santa Maria Times*, May 17, 2010, and letters to the *Los Angeles Times*, July 28, 2010. Paul Daniel Marriott's observations are cited from his *Saving Historic Roads: Design & Policy Guidelines* (1997: Preservation Press).

Engineers spoke about the roadway's conditions in "Future Crowded for Freeways," by Doug Willis, *Los Angeles Times*, January 29, 1984.

Arroyo Fest 2003 was promoted by the Arroyo Seco Foundation. Its 2020 sequel that never happened is detailed in "Open streets event 'Arroyo Fest' to shut down 110 Freeway for one day in 2020," by Blanca Barragan, *Curbed LA*, September 18, 2019.

CHAPTER 2: THE HOLLYWOOD FREEWAY

The union of the Hollywood and Golden State Freeways is sourced from "Hollywood, Golden State Freeways July Weld Set," *Valley Times*, May 4, 1968. The dedication is chronicled in "Hollywood Freeway Section Dedicated; to Open July 10," by Charles Nazarian, *Valley News*, July 4, 1968. That same issue features "Teenage Pilot Uses Freeway To Land Safely."

Much of the curse of Cahuenga Pass is drawn from the *Los Angeles Times*. Cecilia Rasmussen's articles are particularly colorful and informative, including "Curse of the Cahuenga Treasure" in the January 23, 2000, edition. A comprehensive essay on the Pass's buried treasure is "The Cahuenga Pass Treasure," by Joshua Alper, *Southern California Quarterly* (April 1999: Historical Society of Southern California).

There are numerous accounts of the Pacific Electric train that traveled the pass. One key source is "Pacific Electric Tracks in the Hollywood Freeway: A Missed Opportunity," by Ralph Cantos, for the Pacific Electric Railway Historical Society. General Electric produced promotional films in 1946—*Lifestream of the City* and *Way of the Future*, available online—that feature the PE lines running down the freeway median.

Ennis Combes and, in the sidebar, the Basque shepherd, are from Cecilia Rasmussen's aforementioned article.

Campo de Cahuenga background is largely influenced by California State Parks and newspapers, while the nugget about Cecil B. DeMille is from "When the Cahuenga Pass Was Rustic," by Nathan Masters, KCET.org, April 18, 2014.

Descriptions of Hollywoodland are from a 1925 real estate brochure touting the subdivision.

Names of former communities in Los Angeles can be accessed via myriad historical maps in public record.

The Gold Rush's San Fernando Valley shooting location is verified by IMDb. *Four Horsemen* info: "Mission San Fernando Rey de Espana Then and Now," by Eric Mercado, *Los Angeles* magazine website, October 16, 2014.

The *Los Angeles Times* thoroughly recorded the progress of the Hollywood Freeway in the early 1940s. Also "Four Minutes From Bowl To Vineland," *San Fernando Valley Times*, December 19, 1939.

The June 1940 dedication (including the "Old Don Gaspar de Porta" passage) is detailed in "Eight-Lane Cahuenga Pass Freeway, Costing $1,500,000, Dedicated at Colorful Exercises," *Los Angeles Times*, June 15, 1940, and "Civic Rites Will Mark Dedication," *Los Angeles Times*, June 14, 1940.

The abstract "The Fundamental Law of Road Congestion: Evidence from US Cities" was authored by Gilles Duranton and Matthew A. Turner, *American Economic Association*, October 2011.

The Hollywood Anti-Parkway League's grievances are covered in various newspapers, including "Public Money For Freeway Protested," *Valley Times*, August 15, 1940.

Information about the 1870 toll in the Cahuenga Pass (sidebar): www.cppoa.org/images/cahuenga_pass_chrono.pdf

Alternate routes for the freeway skirting Whitley Heights, including proposed maps, are drawn from the *Valley Times*, June 16, 1948. The Hollywood Bowl's gripes are voiced in "Freeway Traffic Faces Debacle," *Valley Times*, January 5, 1948. Gluck's Letter to the Editor: "Valley Forum: Protest

On Freeway," April 13, 1948. Another source documenting 1948: "The Hill, Hollywood's Mediterranean Village," by Patricia Bennett, *Los Angeles Times*, July 12, 1992.

Caltrans's Heinz Heckeroth told me about the thinking behind the freeway's bus line, which is also outlined in a 1953 Transit Study called "Express Busses on Freeways," by the Los Angeles Metropolitan Traffic Association.

The full *Los Angeles Times* article on the Four Level: "This Is Hollywood Freeway Route," August 20, 1948.

Criticism of the *Los Angeles Times* comes from their Letters section, January 7, 1948.

Mascarina Bolluena's saga: "This Is Hollywood Freeway Route," *Los Angeles Times*, August 20, 1948.

In addition to *California Highways* magazines, the State Highway Division's displacement figures are cited in "Homes Bow To March Of Progress," *Valley Times*, September 18, 1950.

Gene Sherman writes about rogue teen drivers on the closed freeway in "Cityside with Gene Sherman," *Los Angeles Times*, April 8, 1954.

The opening of the April 1954 link is drawn from the *Los Angeles Times'* April 15 and April 16 editions, as well as the *Valley Times*.

MGM's movie budget is based on its 1955 production calendar, which is sourced from "The American Film Industry In The Early 1950s," Encyclopedia.com.

The "Nude Outshines Freeway Opening" headline is from the *Colton Courier*, April 16, 1954.

Buried treasure and rotten-egg odors: "Four-Level Interchange Is a Four-Letter Word to Some and a Marvel to Others," by Richard Simon, *Los Angeles Times*, August 18, 1995.

The sidebar about Botts' Dots is informed by *California Highways* and "Rumbling Botts' Dots Are Freeway Lifesavers," by Jennifer Kerr, *Los Angeles Times*, October 6, 1985.

The reader mourning Valley Plaza Park is Mrs. Jerry Scassa, writing to the *Van Nuys News*, January 21, 1968.

The Valley Association for Freeways quotes are attributable to "Freeway Group Backs Engineers," *Los Angeles Times*, July 16, 1954.

Regarding "Freeway Neurosis" (sidebar): It sounds like Grammer was just having a really bad day. Besides fighting traffic, he was stressed about his two children, who had recently been bitten by a dog. "'Freeway Neurosis' Freezes Driver in Rushing Traffic," *Valley News*, July 24, 1955.

In addition to *California Highways*, the extra-wide Highland Avenue ramps are spotlighted in "Crawling Commuters May Fume but There IS an Explanation," by Jack McCurdy, *Los Angeles Times*, June 4, 1961.

Reportage by the *Valley Times* contributes to my passage on the July

1962 freeway extension, as well as the Valley's growth.

The "Downtown-oriented design" scheme is sourced from *Let's Go L.A.*, January 22, 2017. Separately, economic considerations also factored into the lack of some transition ramps, as explored in "Freeway Link to Aid North-South Traffic," by George Garrigues, *Los Angeles Times*, April 25, 1965.

Several eccentric anecdotes are taken from "If It's Unusual It Will Happen on Beat of Freeway Officer '3-FM-26,'" by Jack Schermerhorn, *Los Angeles Times*, March 17, 1965. The Doris Day movie is captured in a photograph by Bruce Cox, *Los Angeles Times*, January 27, 1967. The photo of *Spartacus* shooting near the freeway is viewable online, and its Universal location is confirmed at www.movie-locations.com/movies/s/Spartacus-1960.php.

My conversation about chickens occurred in 2012 with a fellow youth-league father in Toluca Lake, where my son played baseball. Photographer Mike Meadows published a photo of roosters by the freeway on February 18, 1981. It is part of UCLA Library's Special Collections.

Former Caltrans engineer Arturo Salazar provided me with a detailed rundown of Caltrans's "ghost ramps," including the old Barham ramps on the Hollywood Freeway (sidebar).

Kent Twitchell's mural is recapped nicely in "L.A. Given The Brush—But For Colorful Murals and Paintings," Associated Press, *Fort Lauderdale News*, July 17, 1975.

L.A.'s insecurity about not being seen as a world-class city was both my and others' personal observations, inferred from various media portrayals from the 1970s through the 1990s. It was thought that the 1984 Olympics would shed the city of its inferiority complex compared to San Francisco or its Eastern or European brethren, but alas, it did not . . . thus projects like *Steel Cloud*. A good overview of the proposed art project, including mockups, can be found on the blog *Odlandscape*, December 28, 2006.

The proposed park over the 101 is outlined in HollywoodCentralPark.org.

CHAPTER 3: THE HARBOR FREEWAY

Per data produced by CAHighways.org and Caltrans, the Harbor Freeway is one of the top corridors for truck congestion in California. Its reputation as the most dangerous L.A. freeway is additionally cited in "Harbor Freeway Downtown Racks Up the Most Wrecks," by Ronald B. Taylor, *Los Angeles Times*, November 24, 1989.

Phineas Banning, the dredging of San Pedro Harbor, and Los Angeles's annexation of Wilmington and San Pedro are largely sourced from "CityDig: How Los Angeles Annexed the Port on a Shoestring," by Glen Creason, *Los Angeles* magazine website, September 23, 2015, and *Historic Tales from Palos*

Verdes and the South Bay, by Bruce Megowan and Maureen Megowan (2014: The History Press).

Regarding World War II, newspaper accounts from 1941 make the case that a freeway would provide benefits to national defense—the *Long Beach Independent, Pomona Progress-Bulletin*, and *San Fernando Valley Times* among them.

Spruce Goose information (sidebar) is pulled from multiple sources, including History.com.

The April 26, 1945, headlines are derived from the *Wilmington Press*, April 26, 1945. The "big Pacific push" passage is from "Harbor Freeway Contracts Let," *Los Angeles Times*, April 20, 1946.

The Terminal Island Freeway's unenviable claim to "lowest traffic volume" is detailed in "Tear Down a Freeway? In Southern California? It Could Happen," by Hillel Aron, *LA Weekly*, December 28, 2015. The Congress for the New Urbanism is among those advocating its decommission. One advantage of the forsaken freeway is that Hollywood sometimes rents it out for movie car chases. It makes cameos in *Terminator II, To Live and Die in L.A.*, and *Air America*. "Life in the Fast Lanes: A Look at Milestones in Freeway History," by Cecelia Rasmussen, *Los Angeles Times*, February 16, 1994.

The Figueroa Freeway, connecting to the Arroyo Seco Parkway, was envisioned in the early 1940s, per *California Highways* and newspaper accounts.

Various sources are utilized for the rivalry between Long Beach and San Pedro as they vied for a freeway. "Chamber Spurs Harbor Freeway," *San Pedro News-Pilot*, September 15, 1948, is particularly enlightening.

The sidebar about Richard Katz is drawn from "Katz Tries Going With L.A. River's Flow: He sees a stream of cars where no man has dared to drive before," by Bella Stumbo, *Los Angeles Times*, October 15, 1989.

My portrait of Kenneth Hahn is informed by a multitude of publications, including his obit in the *Los Angeles Times* by Richard Simon, Judith Michaelson, and Bettina Boxall, October 13, 1997.

The March 1948 study of displaced buildings: "The Pilot's Log by the Staff," *San Pedro News-Pilot*, May 19, 1951. In addition to *California Highways* issues, displacement stats are from "Rehousing System Sought to Speed Freeway Work," *Los Angeles Times*, March 20, 1948. To give some idea of scope, the Home Builders Institute of Los Angeles estimated that 60,000 people had to relocate due to the building of the Hollywood, Santa Ana, and Harbor Freeways after WWII, per the *Pasadena Star-News*, April 24, 1947.

Hahn's disapproval of the Harbor Freeway route is chronicled by various newspapers and public meetings from 1948 to the early '50s. Associated Press wire services reported on Hahn's decision to hire a blimp in "Harbor Freeway Route Protested," *Wilmington Progress-Bulletin*, March 12, 1948. Hahn's words are pulled from the *Wilmington Press-Journal*: "Ask Re-Survey of Harbor Freeway Route," July 13, 1948, and "Councilman Hahn Calls Proposed Route 'Stupid,'" March 24, 1950.

Assemblyman Thomas's resolution is detailed in "Harbor Freeway Plans Approved," *Pomona Progress-Bulletin*, April 13, 1950. The L.A. City Council vote is recounted in "Harbor Freeway Route Passes 11-2 in Council," *Los Angeles Times*, August 12, 1950.

Updates of the freeway are primarily sourced from the *San Pedro News-Pilot*, as well as "Harbor Freeway Route Approved By All Groups," United Press, *Pomona Progress-Bulletin*, June 23, 1950.

The Harbor's 1952 ribbon-cutting: "Official Ceremonies Open First Harbor Freeway Link," *Los Angeles Times*, July 31, 1952. Hahn's quotes from the day are from the *San Pedro News-Pilot*: "Hahn Likens Harbor Freeway to Gold Streets of Heaven," February 6, 1952.

The eviction of 7,500 Bunker Hill residents was reported by the *Los Angeles Times*: "500,000 Westside Freeway 'Orphans' to Join Frenzy," by Ray Hebert, October 19, 1964. The quotes about Bunker Hill are drawn from *California Highways* archives and the *Los Angeles Times*, mostly "Blighted Bunker Hill Destined to Regain Old Majesty," April 28, 1958.

The sidebar about the unearthing of an 1893 shipping log: "The Pilot's Log By Bynner Martin," *San Pedro News-Pilot*, August 4, 1953.

The miniature golf course saga appears in "The Pilot's Log by The Staff," *San Pedro News-Pilot*, May 19, 1951.

I tapped several sources for Thomas's biography, including "Thomas' Climb to Assembly Credited to Four Policies," by Ed Kenyon, *San Pedro News-Pilot*, October 30, 1956.

Meetings to establish the southern terminus were solidified in 1950. "Final Harbor Freeway Meet Friday," *Wilmington Press-Journal*, August 7, 1950.

The 1954 groundbreaking: "10 Turn Earth to Mark Freeway Start," *San Pedro News-Pilot*, July 23, 1954.

Miss Freeway Link is invoked in "Harbor Freeway Extends Its Reach," *Los Angeles Times*, March 23, 1954 (reprinted March 23, 2006).

Hapless motorist Greg Morton's ordeal ("Motorist Marooned on Freeway Divider—Completely Ignored") is told in the March 8, 1958, issue of the *Los Angeles Times*.

The *Los Angeles Times* details the origin of call boxes on May 14, 1960, in their article "City OKs 14 Emergency Telephones on Freeway." For the sidebar, data for call boxes is drawn from LASafe.net/programs and "Riverside County to remove 225 highway call boxes," by David Downey, *Riverside Press-Enterprise*, March 13, 2019.

Operation Airwatch: "If Trouble's Down Below It, Whirlybird Lets You Know It," by Ben Zinser, *Independent Press-Telegram*, February 2, 1968. Details of Powers's accident mostly pulled from "The Francis Gary Powers Helo

Crash," Check-Six.com, August 1, 1977. As an 11-year-old, I remember watching KNBC anchor Jess Marlow choke up on-air while reporting on the deaths of his colleagues, Powers and George Spears.

The *Daily Independent Journal, Los Angeles Times*, and AP comprehensively document Dodger game congestion in 1958. Rob Shafer's Liberace article appears as "Double Trouble in Traffic" in the July 26, 1959, edition of the *Pasadena Independent Star-News*.

The horrific Harbor Freeway incidents are featured in the *Wilmington Daily Press Journal, Pomona Progress-Bulletin, San Pedro News-Pilot, Long Beach Independent, Los Angeles Times, San Francisco Examiner, Raleigh (N.C.) News and Observer, Van Nuys News, Sacramento Bee*, and *San Bernardino Sun*. The stabbed cabbie is from "Woman Stabs Taxi Driver On Busy Harbor Freeway," United Press International, *Ventura County Star-Free Press*, September 13, 1961.

The lethal stretch south of the Four Level is cited in Caltrans's *Going Places*, July/August 1989. The same issue features Caltrans engineer Ted Roworth's quote about double-decking the freeway.

In the sidebar, Hahn's lobbying for a chain-link median is covered by the *Los Angeles Times*: "Hahn Asks Barriers Be Installed on New Link of Harbor Freeway," August 2, 1959, and "Barriers on Freeway Win County Praise," May 13, 1962.

Various sources account for the Harbor Freeway as the world's busiest, including "Early Opening for Harbor Freeway Link," *Press-Journal*, September 19, 1958. Also "Harbor Freeway Sector Opens With Ceremony," *Los Angeles Times*, September 25, 1958.

My portrait of the September 1962 opening (and Miss Wilmington's photo) is drawn from "Welcome Wilmington Scissors Open Freeway," *San Pedro News-Pilot*, September 27, 1962. Additional news is from "Final Harbor Freeway Link to Be Opened," *Los Angeles Times*, September 24, 1962.

Monorail dreams for the Harbor Freeway—all the way up to Dodger Stadium—started in the early 1950s and continued until the '70s. Key sources include: "Proposal for L.A. Monorail System Told," by Ray Hebert, *Los Angeles Times*, June 16, 1975; "Hahn Urges Monorail to San Pedro," *Los Angeles Times*, October 17, 1961; and "Goodell Monorail" proposal, Los Angeles Metropolitan Transit Authority, September 4, 1962. The *Valley Times* published a proposed map of a metro monorail system under "Detailed Route of Monorail System From Valley To Harbor," February 4, 1954.

Bridge dedication details: "Vincent Thomas Bridge to Be Dedicated Saturday," *Los Angeles Times*, September 27, 1963.

The 1970 completion is previewed in "Freeway dedication set Thursday," *San Pedro News-Pilot*, July 8, 1970. The "Old time autos, pretty girls" headline appears in the *News-Pilot's* July 9, 1970, edition.

The changeover to the 110 route number is expounded by Caltrans and "Harbor Freeway designated Interstate 110," by Sarah Bottorff, Copley News Service, *San Pedro News-Pilot*, July 23, 1982.

Aside from Caltrans, details about the 110's Transitway and double-decking are based on: "Caltrans Proposes Transitway System Along Harbor, Century Freeways," by Doug Smith, *Los Angeles Times*, March 12, 1981; "Planners decide on bus, carpool lanes for Harbor Freeway," by Warren Robak, *San Pedro News-Pilot*, October 10, 1983; "$35-Million Busway Plan for Interstate 15," by Bill Boyarsky, *Los Angeles Times*, February 27, 1984; "Budget shortfall sidetracks Harbor Freeway car-pool lane," *Daily Breeze*, July 28, 1985; and "Changes Urged in L.A. Double-Decking Plan," by Myrna Oliver, *Los Angeles Times*, January 19, 1990.

A multitude of sources peg the Harbor-Century Freeway Interchange at 130 or 132 feet.

The *Los Angeles Times* recaps the tragedy of Daniel V. Jones, most prominently with "Man Kills Self as City Watches," by Alan Abrahamson and Miles Corwin, May 1, 1998. I also witnessed the tragedy on TV as it transpired and accessed old clips on online video platforms.

CHAPTER 4: THE GOLDEN STATE FREEWAY

In addition to *California Highways* archives, Interstate-Guide.com provides useful information on the history of the Golden State Highway and Freeway.

My data for trucks comprising 25 percent of the traffic at the L.A./Kern County line is deduced from a 2018 study by the Office of System Performance, Traffic Data Branch, Division of Traffic Operations.

The "backbone" quote is attributed to Caltrans District 7 Director Carrie Bowen. "Caltrans holds ribbon cutting for new I-5 truck lane and freeway widening in Santa Clarita," by Anna Chen, *The Source*, Metro.com, December 9, 2014.

My interviews with Caltrans engineer Heinz Heckeroth informed much of the East L.A. Interchange section, as did "Highway Recollections of Heinz Heckeroth: Oral History Interview: October 29, 2003," which resides in Caltrans's Transportation Library. The *Redlands Daily Facts* terms the interchange "Octopus" ("Major freeway change planned for Octopus") in its August 23, 1965, issue. Bill Keene's myriad nicknames for traffic and junctions are well-documented, and I recall them from KNX traffic reports. Downtown's Four Level is officially the Bill Keene Memorial Interchange, posthumously named in 2004. It's fun to read the state Senate's Legislative Counsel's Digest endorsement of the measure, which argues that "Mr. Keene made traffic reports more interesting by referring to accidents with words like 'cattywampus,' 'chrome cruncher,' and 'paint peeler.'"

Two valuable sources for U.S. 99 are "Historical Tour of US 99 Golden State Highway," by Michael F. Ballard, *Southern California Regional Rocks and Roads* and "Historic US 99 Guide—The Ridge Route," GBCnet.com/ushighways.

In the sidebar, freeway and expressway differences are delineated on several state department of transportation sites and on CAHighways.org.

Numerous sources contribute to my portrait of Griffith J. Griffith, including "Happy 120th Griffith Park, but your founder was a jerk," by Robert Petersen, *Off-Ramp*, KPCC, March 23, 2016.

There are various accounts from 1955 of the fight to save Griffith Park, including "Fight Intensified On Park Freeway," *Los Angeles Times*, March 6, 1955. The loss of various recreational facilities is pulled from "City Opposes Golden State Freeway Route," *Valley Times*, March 26, 1954.

Mayor Poulson's reroute request: "Poulson Accuses State Of Land Grab For Freeway," United Press, *Wilmington Daily Press-Journal*, March 26, 1956.

The Hollenbeck Park section is informed by "The prettiest park in Los Angeles and why a freeway runs through it," by Carolina A. Miranda, *Los Angeles Times*, August 12, 2015, and *The Folklore of the Freeway: Race and Revolt in the Modernist City*, by Eric Avila (2014: University of Minnesota Press).

Boyle Heights' history is drawn from several sources, including "Boyle Heights, the land of freeways," by Avishay Artsy, KCRW, October 6, 2015, and my interview with Pastor Stephen "Cue" Jn-Marie. Councilman Edward Roybal's quote: "100 Meet to Protest Proposed Freeway Link," *Los Angeles Times*, October 28, 1953.

Division of Highway quotes about East Los Angeles are attributable to the abstract *Paving East Los Angeles: Building the Freeway System in East Los Angeles, 1940–1970*, by Dylan Warren Logas, CSU Long Beach, April 11–13, 2019. Land values are derived from "The Historical Roots of Gentrification in Boyle Heights," by Gilbert Estrada, KCET.org, September 13, 2017.

California Highways covers the viaduct through Hollenbeck (sidebar).

There are various accounts of the "Battle of the Bulge," primarily from the *Valley Times*: "New Fight Looms on Freeway 'Loop,'" November 20, 1953, as well the paper's editorial dated May 29, 1954: "Valley United Front Could Expedite Freeway Building."

A portrait of Van Griffith appears in "Architectural Rambling—The Van Griffith Estate," *Daily Mirror*, June 22, 2011. Mayor Bowron's insistence that Griffith resign is taken from "L.A. Mayor Calls For Griffith Resignation," UP, July 23, 1946.

Van Griffith's lawsuit and his battle with Frank Durkee are conveyed in the *Los Angeles Times* and *Valley Times* in the period covering January 1955 through April 1955. Larry Norman's plea to Durkee: "Call for End of Freeway

Work in Park," *Valley News*, November 13, 1955.

Beyond *California Highways* reports, the 1955 opening is sourced from "Open Section of Freeway North of San Fernando," *Valley News*, July 12, 1955, and "New freeway link ends traffic jams," *Newhall Signal*, September 8, 1955.

Van Griffith losing his appeal: "Deny Injunction to Halt Freeway," *Valley Times*, September 27, 1955.

David Jordan shares his outrage with the *Valley Times* in "Raps Council for Freeway Row Action," April 1, 1956.

The "unused small hill" is mentioned in "Griffith Park Ruling Asked," *Valley Times*, May 24, 1955. The subsequent "Battle of the Barricades" was picked up by local papers in January of 1956. My sources are primarily the *Los Angeles Times* and *Valley Times*, whose "Griffith Park Barricade Halts Valley Freeway" is from January 6, 1955.

The saga between the state, parks, and L.A. City Council has a plethora of sources, mostly: "Council Blasts at Barriers to Freeway Work," *Los Angeles Times*, January 10, 1956; "Park Barriers Removed in Freeway Dispute," *Los Angeles Times*, January 24, 1956; "Barricade Battle Ends With State Victorious," *Long Beach Independent*, January 24, 1956; and "Park Board Asks State to Change Freeway Grading," *Valley Times*, February 3, 1956.

The $1 million settlement is enumerated in "Griffith Park R.R. to Reopen," *San Pedro News-Pilot*, July 5, 1957.

The teed-off golfers come courtesy of the *Valley Times*: "Work Begins on Freeway; Golf Suffers," August 14, 1956.

In the sidebar, the Los Feliz Drive-In address appears in movie ads in the *Los Angeles Times*, March 15, 1956. Former Los Angeles Councilman Tom LaBonge shared with me his days growing up in the Los Feliz area during various interactions.

The preview of the freeway is drawn from "Link Opens on Golden State Freeway Project," *Los Angeles Times*, May 15, 1957.

Man forced to move twice due to freeways: "Abuse Piled on Freeway Negotiators," *Los Angeles Times*, April 8, 1957.

Lomie Puckett's stand-off with the highway folks went national, reported by wire services. The featured headlines and Puckett's quotes are from the *Redlands Daily*, August 5, 1958. The *Tampa Tribune* article is entitled "Heartaches On Superhighways," by Tom Henshaw, June 7, 1959. Puckett's court case is largely sourced from "Gun Defender of Freeway House in Court," *Los Angeles Times*, October 22, 1958. Ross Puckett's new job is revealed by the Associated Press and *Los Angeles Times* ("Couldn't Lick 'Em, Puckett Son Joins 'Em"), both on September 30, 1958.

Besides *California Highways* maps, a rendering of the proposed Chavez Ravine Freeway (sidebar) appears alongside "Drive Launched for Chavez

Ball Park," by Paul Zimmerman, *Los Angeles Times*, March 12, 1957.

The November 1963 ribbon-cutting is chronicled in "San Fernando Plans Set for Freeway Fete," by Porter Flint, *Valley News*, October 27, 1963. The same newspaper displays a "Tom Carrell's Fabulous Freeway Party" advertisement on November 8, 1963. The drop-off in traffic was observed as early as 1955 with the opening of the western portal: "Freeway Deadens Once Teeming Rd.," *Valley Times*, August 11, 1955. Separately, traffic studies have pointed out that although volume on surface streets often dipped precipitously when nearby freeways opened, it usually swelled again over time as the population grew (a small consolation for the businesses that had to close due to the "dead" period).

Data comparing the 5 and 101 Freeways: "Traffic Experts Report Easing of Bottlenecks," by Art Ryan, *Los Angeles Times*, June 24, 1962.

Various sources cover the conversion of the 5's expressway sections to freeways throughout the 1960s, including "Southland Roads to Get $164 Million," *Los Angeles Times*, November 4, 1962. Heckeroth also spoke to me extensively about the Ridge Route. Defying convention, it was his idea to have the northbound lanes switch to the left side of the freeway north of Castaic to take advantage of the more gradual incline for trucks. He also instigated the creation of Truck Lanes through Newhall Pass.

Accounts of Newhall Pass's dangerous traffic patterns are sourced from "Newhall's Killer Roads: 900 Dead or Injured," by Mary Altschul, *Los Angeles Times*, September 11, 1969, and "If the Big Rigs Won't Slow Down Throw the Bastards Into Jail," *Newhall Signal*, April 17, 1974.

Two weird accidents (sidebar): "Motorists Lose Bit of Everything on Freeways: TVs, Cash, Rings," *Los Angeles Times*, November 17, 1968, and "Tragic Body-on-Freeway-Sign Story Grows Older," by Neal Broverman, *Los Angeles* magazine website, November 2, 2015, the latter one of several sources on Richard Pananian's death.

Besides *California Highways*, information on the Golden State-Antelope Valley Interchange is derived from "Freeway Building to Resume," *Newhall Signal*, December 27, 1972, and "'Quake' interchange finish years away," United Press International, *Tulare Advance-Register*, August 30, 1971. The 150-foot height of the interchange is cited in "Work on Quake-Hit Freeway Stops," UPI, *Sacramento Bee*, August 28, 1971 (however, I do tend to question the article's accuracy of its height and was unable to independently verify it from other parties). The Harbor-Century (Judge Pregerson) Interchange, at 130 feet, was billed "the tallest" in California when it opened in 1993.

Local and national media and wire services reported on the '71 Sylmar Quake. Details of the crushed pickup occupants are drawn from "Two Killed In Collapse of Overpass," *Valley News*, February 11, 1971. "Calif. Bridgebuilding ambitious, disastrous," by Mary Neiswender, *Long Beach Press-Telegram*,

October 19, 1972, recounts the damaged bridges.

Destruction of the Golden State-Antelope Valley Interchange is sourced from *California Highways*, SoCalRegion.com, and also "Highway Damage: $30 Million," Copley News Service, *Redondo Reflex*, February 17, 1971. Councilman Donald Lorenzen is quoted in "Damaged Valley Interchanges Called Major Planning Folly," by Irv Burleigh, *Los Angeles Times*, February 11, 1971.

The *Newhall Signal* documents the April 1974 completion work in "Tin Lizzies On The Freeway," April 29, 1974.

The 1994 Northridge Quake was extensively covered. Freeway damage is primarily drawn from "Earthquake, The Long Road Back," *Los Angeles Times*, January 28, 1994. Clarence Wayne Dean's accident is addressed in "L.A. Dodges Big One, But It's Bad Enough," by Tracey Kaplan and Marc Lacey (*L.A. Times* reporters), *Salt Lake Tribune*, January 18, 1994. Shockingly, a second motorist soared off the same severed freeway connector just days later in broad daylight, despite barriers warning that the bridge was out. Even more amazing, the motorist, Tony MouFarrege, escaped serious injury. Like something out of a silent movie, his van "dropped like a bomb into [a] bulldozer scoop and sat there, tail in the air." "Driver Survives Dive Off Freeway Overpass," by Jeff Schnaufer, *Los Angeles Times*, January 21, 1994.

I accessed government documents for bridge damages from an engineering standpoint. The United States Department of Commerce Technology Administration was helpful in this regard, particularly their report *1994 Northridge Earthquake: Performance of Structures, Lifelines, and Fire Protection Services*.

Public records from the California Legislature, and Caltrans press releases, provided information on the resolution to institute the Fallen Workers Memorial Interchange.

The Gavin Canyon restoration is sourced from "Golden State Freeway Opens 3 Weeks Early," by Henry Chu, *Los Angeles Times*, May 18, 1994. The Newhall Pass restoration was informed by "Reopening of Freeway Interchange Celebrated," also by Henry Chu, *Los Angeles Times*, November 5, 1994.

In addition to *California Highways* and other sources, the Los Angeles Almanac Online explored the etymology of the Grapevine.

CHAPTER 5: THE FOOTHILL FREEWAY

The *Los Angeles Times'* harping about traffic near Devil's Gate is from "Solid Footing Found for Devil's Gate Dam Span," December 14, 1952. As for the official October 1955 unveiling, the bridge actually opened to two-way traffic on August 3, 1955, but the northern buttress still needed two months of work before all lanes were functional. "Devil's Gate Section of Freeway is Open," *Pasadena Independent*, August 4, 1955.

Speaking of interstate numbers (sidebar), both I-710 and I-110 are not technically alternate routes to mother freeways. But per Caltrans deputy district director Chuck O'Connell, their definitions were loosened because they serve "interstate transportation of goods from the ports of Los Angeles and Long Beach." "Honk if You Love Answers to Your Questions," by Hugo Martin, *Los Angeles Times*, April 30, 2002.

The $14 million patch from I-5 to Hansen Dam: "State OK's Segment of Foothill Freeway," *Los Angeles Times*, April 13, 1958.

Numerous late '50s articles from the *Valley Times* and *Los Angeles Times* cover opposition in La Cañada, Montrose, and more affluent areas. Assemblyman Lanterman expresses his views in "Foothill Freeway Bill Withdrawn," by Don Snyder, *Los Angeles Times*, June 11, 1965.

In the sidebar, the merging of La Cañada and Flintridge is primarily sourced from the City of La Cañada Flintridge website. The Foothill Freeway actually had a hand in it. As one resident, Warren Hillgren, put it, the county "had recently divided our community in two, with a freeway decision in which we had no participation—something we seek should never happen again" by having a larger population base.

Glendale Mayor Barnes's rebuke against Warren Dorn appears in "Glendale Fights Lonesome Battle Favoring Foothill Freeway Route," *Los Angeles Times*, September 11, 1964.

Warren Dorn biographical information is drawn from various obituaries in 2006.

The "Multi-Million Dollar Question" *Los Angeles Times* headline fronts an article by Don Snyder, May 31, 1964.

The loss of the La Cañada school: "La Cañada School Gets OK to Move," *Pasadena Independent*, July 26, 1968.

In the sidebar, the figures cited (as of September 11, 1964) for the number of units acquired and threatened: "Glendale Fights Lonesome Battle Favoring Foothill Freeway Route," *Los Angeles Times*. The tunnel scenario is sourced from "Editorial: Pull the plug on the 710 tunnel," *Los Angeles Times*, May 23, 2017, and "State Route 710 North Study: Freeway Tunnel Alternative Fact Sheet," a Metro/Caltrans Report.

The photo of Councilman Nowell christening the groundbreaking is from the *Los Angeles Times*, May 25, 1968.

Lanterman is quoted in the *South Pasadena Review*: "Freeway Plans, Foothill Route Criticized by Lanterman," December 23, 1968.

All January 1969 storm details are sourced from the *Los Angeles Times* and *Valley Times*. "Flood Havoc In Southland: Northeast Valley Area 'Desolated'" headline is from the *Valley Times*, February 25, 1969.

My portrait of the bridge collapse in 1972 is drawn by various dailies, most prominently: "'Bridge unsafe' victim of Pasadena tragedy said," *Ukiah*

Daily Journal, October 19, 1972; "In 1972, Foothill Freeway bridge collapses while under construction," *La Cañada Valley Sun*, October 17, 2012; and "Calif. Bridgebuilding most ambitious, disastrous in U.S.," by Mary Neiswender, *Long Beach Press-Telegram*, October 19, 1972. The state engineer is quoted in "Builder, Safety Officials Hit in Bridge Collapse," by David Rosenzweig, *Los Angeles Times*, November 3, 1972.

Ray Hebert reports on the freeway opening between La Cañada and Pasadena for the *Los Angeles Times* ("Strip of Controversial Foothill Freeway Opens") on July 18, 1972.

Nowell's proposal after the 1969 floods: "Nowell in New Stand on Tujunga," by Michael Wyma, *Van Nuys News*, February 16, 1973.

EIR updates throughout the 1970s are drawn from various dailies and public records.

James Stanley ("so-called environmentalists") is quoted in "More Delays to Snag Foothill Freeway Link," by Irv Burleigh, *Los Angeles Times*, November 27, 1972.

Nowell's gravel interests are outlined in "Work to Begin on Embattled 'Missing Link,'" by James Quinn, *Los Angeles Times*, July 23, 1978. His cozy relationship with developers is well-covered by news of the day, recapped in a July 24, 2009, obituary by the *Los Angeles Times'* Valerie J. Nelson. Jerry and Debbie Decter's crusade against Nowell is drawn from "Oust-Nowell Campaign Falters," by Kenneth Reich, *Los Angeles Times*, February 13, 1973, and "Activist helped bring down L.A. councilman," by Valerie J. Nelson, *Los Angeles Times*, July 27, 2009.

The feared "smog trap" by the Air Pollution Control District is cited in "Freeway Progress Checked by High Level Smog Report," by Irv Burleigh, *Los Angeles Times*, July 25, 1974. The antiquated phrase "smog trap," incidentally, was commonly adduced in many arguments against freeways in the 1970s and 1980s.

The "Wanna Bet?" headline: *Los Angeles Times*, March 11, 1976.

My accounts of Nowell's cynicism while opening a portion of the Simi Valley Freeway are sourced from the *Los Angeles Times*: "New Delay on Simi Freeway Draws Anger," by Ken Lubas, September 23, 1976, and "Freeway Link to Nowhere May Go Somewhere by Next Spring," by Don Snyder, July 25, 1976. Constituent James Moran is quoted in "City Plan for Wash Pushed Past 3 Others," by the *Los Angeles Times'* Irv Burleigh, April 7, 1974.

In the sidebar, Greenpeace's open-water defiance against whaling ships in the 1970s was international news. Anti-710 Freeway activists are portrayed in "Road Work Appeal," by William McPhillips, *Los Angeles Times*, February 8, 1976.

Voices against Nowell: "Letters to the Times," *Los Angeles Times*, March 12, 1976.

Cecilia Rasmussen's *Los Angeles Times* piece from February 16, 1994 ("Life in the Fast Lanes: A Look at Milestones in Freeway History"), aided my research for shoots on closed freeways. The detail about Columbia Pictures renting the Foothill Freeway is from *The Warrior in Me* by ex-cop D. E. Gray (2010: Xlibris). A good primer on *CHiPs* is "*CHiPs* Turns 40: A Guide to the Series' Most Notable (and Easy to Find) Locations," by Todd Munson, *Medium*, September 14, 2017. *CHiPs* episodes, or portions thereof, are also available online; plot recaps, on IMDb.

The sidebar about the CHP taking over freeway patrols is sourced from "CHP Takes Over F'ways," *Los Angeles Times*, October 2, 1969. The *CHiPs* headquarters is my own personal observation and confirmed by "Highway Patrol Officers in Middle of the Action," by Bob Pool, *Los Angeles Times*, May 22, 2003.

A preview of the 1981 dedication was rolled out by the *Los Angeles Times* ("Foothill Freeway Dedication Slated") on March 29, 1981. Kevin Roderick writes about the bridges over Tujunga Wash in "Caltrans Study of Freeway Link OKd," *Los Angeles Times*, April 3, 1977.

The $700 million figure for the freeway: "Tax Measures Fuel a Quiet Freeway Construction Boom," by Mark A. Stein, *Los Angeles Times*, March 22, 1992.

The Rodney King saga was widely reported. Much of my account is pulled from the *Los Angeles Times*. The paper's Richard Serrano penned "LAPD Officers Reportedly Taunted King in Hospital," March 23, 1991. Mayor Bradley's statement is posted in the *Times* under "Bradley Blasts 'Bigotry' of Police Officers," March 20, 1991.

The proposal to nix the 710 extension is well-chronicled by numerous blogs and outlets, including "Caltrans effectively kills 710 Freeway extension after decades-long battle," by Carol Cormaci and Laura J. Nelson, *Los Angeles Times*, November 29, 2018.

CHAPTER 6: THE VENTURA FREEWAY

The San Fernando Valley Historical Society provides an overview for much of this chapter.

Accounts of Harold Bayly and early routing issues through the west Valley are drawn from daily newspapers, including: "Freeway Route Protest Heard," Associated Press, *Pomona Progress-Bulletin*, August 21, 1953; "Protests Over Route of Ventura Freeway Heard," *Los Angeles Times*, December 3, 1953; "Old Ranch Slumbers in Aerospace Clamor," *Los Angeles Times*, June 23, 1963; and "Horse Ranch Sale to Mark Era's End," by Doug Smith, *Los Angeles Times*, May 13, 1985.

Celebrity ranches are profiled in "Edward Everett Horton's Encino

Ranch Estate and the 101 Freeway; How A Celebrity Lost His Ranch to Suburbanization," by Marty McFly, *San Fernando Valley* blog, April 4, 2012.

The name-change to Tarzana is culled from 1928 records, including "Runnymede Section Takes Name Tarzana," *Los Angeles Times*, July 26, 1928.

The headline "State Stands by Route for Ventura Freeway," is from the *Pomona Progress-Bulletin* (article from AP), December 19, 1953.

The state's policy to convert accident-prone stretches of highway (here, U.S. 101) into freeways was instituted in 1955 with both the first leg of the Golden State Freeway in the Valley, and the first leg of the Foothill Freeway where it spanned the dangerous Arroyo Seco at Devil's Gate Dam. "Freeway Ordered at Montecito to Cut Accidents," *Los Angeles Times*, July 24, 1951.

Population post-WWII sourced from Demographia, Los Angeles Community Areas Population & Density: 1950–2010.

Governor Brown's bill authorizing El Camino Real (sidebar): "Valley Lifeline Follows Path Taken By Explorers," by the *Los Angeles Times*' research team, August 6, 1995.

Two significant sources inform the extension of the Ventura Freeway as SR-134: "Plan Joint Hearing On Valley Freeways," *Valley Times*, March 3, 1954, and "Valley Home Owners Battle State Over Proposed Riverside Freeway Route," *Valley Times*, February 12, 1954.

In addition to *California Highways* archives, the *Los Angeles Times* detailed the release of state funds to the freeway: "Ventura Blvd. Traffic Relief Seen by 1959," April 7, 1957.

In the sidebar, the official adoptions of the Ventura Freeway name are covered in "Revised Freeway Program Adopted," May 27, 1955, and "Ventura Freeway Named," AP, *Redlands Daily Facts*, February 5, 1957.

Accounts of the freeway through Woodland Hills are mostly drawn from "Future Need for Freeways," by Bob Halle, *Citizen-News*, April 6, 1960, and "Ventura Freeway Acclaimed By All," *Valley Times*, April 6, 1960.

The *Valley Times* declared Ventura the fastest-growing county in the U.S.: "Residents of Valley Invited to See Edgemont Homes," September 12, 1960.

The "fraying strands of innocence" stories are culled from: "Steer Disrupt Freeway Traffic," *Valley Times*, October 16, 1963; "Youth's Dash Down Freeway All in Vain," by Sid Bernstein, *Los Angeles Times*, September 7, 1963; and "Plane, Short of Gas, Drives Along Freeway," *Ventura County Star*, October 20, 1962.

Adam West's auto mishap (sidebar): "Crash...Bam...Pow," *Los Angeles Times*, February 13, 1969.

The early opening of the Ventura-San Diego Interchange is sourced from "Sections of 2 Freeways to Be Opened Thursday," *Los Angeles Times*,

July 1, 1958. The caption "It's All Yours" and accompanying photograph are from the *Van Nuys News*, July 8, 1958.

California Highways promoted the eighty-mile uninterrupted run to San Juan Capistrano, as did the *Los Angeles Times*: "Freeway in Valley to Open April 5," March 13, 1960.

Governor Brown's ribbon-cutting snafu is portrayed in "Because of Clippers, A Freeway Is Dedicated," *Valley Times*, September 22, 1962.

My accounts of the odd Ventura-Golden State Interchange, and its goat-nibbling guest, are drawn from: "New Interchange to Link Two Valleys," *Los Angeles Times*, June 24, 1966; "Goat to Star at Freeway Opening," by Ken Fanucchi, *Los Angeles Times*, August 24, 1967; and "Burbank Fete Will Hail Dedication of Freeways," *Valley Times*, August 21, 1967.

There are a number of sources for the construction of the Colorado Street Bridge. For its (freeway) replacement, I rely on "Arroyo Seco Span Cement Job Starts," *Los Angeles Times*, August 7, 1951. Suicide stats are from "New Pasadena Bridge May Be Completed Early," *Los Angeles Times*, May 25, 1953. The Pioneers Bridge (officially the Pasadena Pioneers Bridge) info is taken from a plaque on-site.

The Colorado Freeway was approved by the state and Los Angeles City Council on April 17, 1952. "City Council Approves Colorado Freeway Pact," by Harry J. Frawley, *Valley Times*, April 17, 1952. "Colorado Freeway Open Fri.," *Valley Times*, June 23, 1954, covers its opening. Caltrans's Arturo Salazar shared with me the vestiges of the Colorado Freeway.

The 1971 Eagle Rock fiasco is spelled out in "Ventura Freeway Criticized at Dedication," by Don Snyder, *Los Angeles Times*, August 19, 1971. Arthur Snyder's appeal to Governor Reagan: "Faster Action Sought From State on Landscaping New Freeways," *Los Angeles Times*, May 28, 1972. The scathing *Times* headline "Ventura Freeway Criticized at Dedication" is from August 19, 1971, with Don Snyder in the byline.

The Noise Abatement Team in the sidebar is sourced from "Watch Out, the CHP May Be Listening," by Mark Forster, *Los Angeles Times*, August 28, 1977.

The wheelchair lift problem on RTD buses was outlined in the bus agency's own newsletter, *RTD Headway* ("Mechanical problems cripple accessible service"), March 1981.

Engineering flaws of the 101/405 intersection are covered in "Interchange Relief Coming . . . Later," *Los Angeles Times*, August 6, 1965. Also "Freeway Mistakes Will Cost," by Phil Hanna, *Valley Times*, August 5, 1965. Progress (or lack thereof) of the interstate's renovations is conveyed in "San Diego, Ventura Freeway Bottleneck," by Harold N. Hubbard, *Valley Times*, August 17, 1967, and "Sick Valley Interchange May Never Get Well," by Martha Willman, *Los Angeles Times*, November 21, 1971, from which the first paragraph is reprinted.

Alweg's proposal is sourced from "Monorail Offer Is Premature, MTA Aides Say," *Valley Times Today*, April 1, 1963.

The *Los Angeles Times*, *Valley News*, and *Valley Times* published public forums of the mid-1960s, including dissension about the proposed Whitnall Freeway. Opposition to double-deck freeways from neighbors and Supervisor Warren Dorn are from "Valley Leaders Oppose Double-Deck Freeways," by Gordon Grant, *Los Angeles Times*, November 26, 1967. Interestingly, Dorn's "no" vote for double-stacked freeways didn't stop him from pursuing monorail lines above freeways, traveling along the center medians: "Dorn Proposes Monorail Above Freeways," by Ray Hebert, *Los Angeles Times*, July 16, 1975. Fellow Supervisor Frank G. Bonelli's advocacy for a second level is expressed in "Double Decks on Freeways Urged Again," *Valley News*, November 19, 1968.

Norman Greene's monorail scheme is covered in monorail blogs, as well as "Monorail Advocate Acclaims His System," *Los Angeles Times*, September 15, 1968. Harry Bernstein's system: "Rapid Transit Dream: Own Monorail Car," by Paul Houston, *Los Angeles Times*, July 2, 1970.

In the sidebar, the consistent use of "Do Not Enter" and "Wrong Way" signage is mentioned in *California Highways* and "Ramp Signs Warn Wrong-Way Drivers," *Los Angeles Times*, August 22, 1965. As one might expect, the first "Do Not Enter" signs appeared on the Arroyo Seco. "Engineers Map Freeway Safety," *Los Angeles Times*, December 15, 1941.

Several sources cover the September 1974 interchange opening in Pasadena, including "Section of Freeway Interchange Slated for Opening in Pasadena," *Los Angeles Times*, September 22, 1974. Caltrans records show the final touches being completed in 1975.

The band America speaks about the origins of their song in "Behind the Song: 'Ventura Highway' by America," by Paul Zolla, *American Songwriter*, May 2020.

"Ventura Freeway—It's Now No. 1" was penned by Ray Hebert for the *Los Angeles Times*, March 10, 1985.

Information about the Adopt-A-Highway sidebar is drawn primarily from "Fast Facts: Adopt-a-Highway," by Darrell Satzman, *Los Angeles Times*, September 1, 1996, and "Roadside Rubbish: Adopt-a-Highway volunteers are putting a dent in trash, but their task is Sisyphean. Drivers need to take part too," by Richard Kahlenberg, *Los Angeles Times*, February 20, 1992.

I tapped several sources for the Obama Highway designation, principally: "'Obama Freeway' resolution OK'd by state Senate," by Taryn Luna, *Sacramento Bee*, May 20, 2017; "Proposal would name freeway after Obama," by Jeff Landa, *Los Angeles Times*, December 26, 2016; and "Ten Things You May Not Know About Eagle Rock," by Tim Loc, *LAist*, March 16, 2017.

CHAPTER 7: THE SAN DIEGO FREEWAY

"Longest" and "costliest" freeway sourced by *California Highways* and "13-Year Job Completed on San Diego Freeway," by Bob Sanders, *Long Beach Independent Press-Telegram*, January 12, 1969.

In the June 7, 2012, issue of *LA Weekly*, 54 percent of respondents said the 405 had the worst traffic, making it "the most-loathed freeway in L.A." per writer Dennis Romero. Bluntly, Angelenos don't need a poll to know this.

California Highways is sourced for the southern terminus of the San Diego Freeway (sidebar), as is the South Bay Historical Society.

Sepulveda Tunnel excavation: "Ask Immediate Sepulveda Start," *Van Nuys News*, July 24, 1928.

Two comprehensive overviews of Sepulveda Pass include "405 Freeway's path tells a story of near-constant change," by Mike Anton, *Los Angeles Times*, July 14, 2011, and "How Sepulveda Canyon Became the 405," by Nathan Masters, KCET.org, June 27, 2017.

A future freeway through the pass is explored in "Planning Great Sepulveda Freeway," *Van Nuys News*, October 26, 1943.

Heinz Heckeroth told me about the financial mechanism known as "Chapter 20 Money."

Relocation figures are garnered from various sources, including "Knight Reveals Speed-up Of Valley Freeway Plan," *Valley Times*, September 21, 1954.

Accounts of the Sunset Boulevard groundbreaking: "Ground Broken for Sepulveda Freeway Unit," *Los Angeles Times*, September 21, 1954.

The name conversion from "Sepulveda Freeway" is covered in "New Names Given to 4 Freeways," *Los Angeles Times*, November 25, 1954.

The accidental comingling of gas and sewer lines dominated newspapers for a few days. My accounts are drawn from: "Fuel Leak Perils L.A.," Associated Press, *San Pedro News-Pilot*; "Mayor Probes Gasline Break," *Mirror-News*, February 28, 1956; and "Fuel Leak Perils L.A.," AP, *San Pedro News-Pilot*.

The juvie car thief: "Policeman Shoots Boy Fleeing Scene of Crash," *Los Angeles Times*, March 29, 1957.

The transition from black to green freeway signs is extensively covered in Caltrans's records, in particularly the March/April 1962 issue of *California Highways*.

March 1957 dedication: "Knight, Poulson Open First Link of Freeway," *Los Angeles Times*, March 30, 1957.

Big Cut details are largely drawn from *California Highways*, as well as the *Los Angeles Times* and *Valley Times* (1960–1962). Valley residents claiming "ocean breezes" are profiled in "Experts Discount Ventilating Value of Sepulveda Pass Cut," *Los Angeles Times*, August 6, 1961.

Congestion around LAX: "L.A. Delegation Urges Construction Speed-up On San Diego Freeway," AP, *Pomona Press-Bulletin*, September 1, 1960.

Accounts of the 1961–1962 storm and mudslides are taken primarily from the *Los Angeles Times*: "Delay Seen in Freeway Completion," by Bill Burns, January 21, 1962, and "Homes Being Removed From Freeway's Banks," April 29, 1962.

The 405's preparation ("lace on the dress"): "Construction on Freeways to Run Until '69," by Al Asermely, *Los Angeles Times*, September 30, 1962.

Much of the December 1962 dedication is captured on film—*A Special Day*, by the California Department of Public Works, December 1961. Other accounts, including Yorty's ribbon-cutting stunts, are drawn from the *Valley Times'* "F'way Link Causes Jam," December 22, 1962, and two *Los Angeles Times* articles: "Memory Lanes," by Mike Anton, July 14, 2011, and "Sigalert and Tieup Mar Freeway Link Opening," December 22, 1962. To be technical about it, speedster Ronald P. Tamkin was only doing 72 in a 65-mph zone. But . . . he was driving home with his newborn son in the backseat.

History has not been kind to Mayor Sam Yorty, as various publications can attest. His poor ranking is cited from Melvin Hollie's book *The American Mayor: The Best & The Worst Big-City Leaders* (1999: Penn State University Press). In the sidebar, Councilman Karl Rundberg's criticism of Yorty appears in: "Rundberg Says Yorty Pulls Phony Publicity," *Los Angeles Times*, March 19, 1963; "Yorty's Helicopter Flights Criticized," *Valley Times Today*, March 19, 1963; and "Rundberg Says Yorty Pulls Phony Publicity," *Los Angeles Times*, March 19, 1963.

Matt Weinstock's column ("Where Traffic Used to Roar—All Is Quiet") appears in the *Los Angeles Times*, March 6, 1963. Marvin Wilson's Royal View Estates ads are from the January 31, 1963, edition.

The 427 homes demolished near LAX: "East-West Freeway Route Selection Near," *Los Angeles Times*, February 23, 1964.

Photos and copy of the picketing mothers and children are sourced from "Lack of Children's Overpass Protested," *Los Angeles Times*, June 25, 1963, and "Picketing Mothers Win New Freeway Support," *Los Angeles Times*, April 7, 1963.

The official San Fernando ribbon-cutting for the northern terminus of the freeway occurred on April 19, 1963 (sidebar). Lieutenant Governor Glenn Anderson, who would later get a freeway named after him (aka the Century Freeway), did the honors: "8.4-Mile Link Opened on San Diego Freeway," *Los Angeles Times*, April 20, 1963.

The blessing by a priest: "Freeway Link Open To Cars; Jam At Rites," *Valley Times Today*, April 20, 1963. The Eden Memorial lawsuit over eleven acres: "Cemetery Assn. Wins Reversal in Its Battle Against Two Freeways," *Los Angeles Times*, December 25, 1962. As occurred with San Fernando Road when I-5 opened, Sepulveda Boulevard saw similar drop-offs in business traffic when the 405 reached the Valley in early 1963, though many merchants welcomed the fact that trucks were now using the freeway, making it easier for folks in autos to shop. "Traffic Less, Business Good on Sepulveda," by Jerry Custis, *Valley Times Today*, May 21, 1963.

UCLA students storming the freeway is covered, among others, in "USC Gets Bid to Rose Bowl; UCLA Students Run Wild," by Phil Fradkin and Dick Main, *Los Angeles Times*, November 22, 1966.

Various dailies report on the Orange County extension, including "Dedicate San Diego Freeway," by Bob Sanders, *Long Beach Independent Press-Telegram*, December 7, 1968, and "Horsemen Help to Dedicate San Diego Freeway's Final Link," *Los Angeles Times*, December 7, 1968.

"San Diego Freeway Now Reality" headline is from the *Los Angeles Times*, December 9, 1968. Article by Ray Hebert.

As usual, price tags for freeway work vary depending on sources; figures for the completed San Diego Freeway here are drawn from the aforementioned "13-Year Job Completed on San Diego Freeway."

Comparisons between the South Coast and San Fernando are posed by David Shaw in "Beauty or Blight Challenge Faces the South Coast," *Los Angeles Times*, December 29, 1968. The *Times* waxes about freeways—and varicose veins—in "Beauty or Blight Challenge Faces the South Coast," also by David Shaw, December 29, 1968.

I vividly recall several trips to Lion Country Safari as a child, including the "Trespassers Will Be Eaten" sign. Accounts of the park's travails were regular news. Incidents in this sidebar are drawn from: "Family of Boy Mauled by Tiger Files Suit," by Jerry Hicks, *Los Angeles Times*, December 1, 1982; "Child in serious condition after being mauled by tiger," AP, *Californian*, October 25, 1982; "Bubbles Takes Lunch Break," by Gordon Grant, *Los Angeles Times*, March 2, 1978; "The sad demise of Bubbles the hippo," AP, *Berkeley Gazette*, March 11, 1978; "Elephant kills game warden at Lion Country," AP, *The Sun*, July 25, 1983; "Elephant Goes Berserk, Kills Park Zoologist," by Evan Maxwell and Bill Billiter, *Los Angeles Times*, July 25, 1983.

United Press International's initial erroneous report on the location of the two crushed motorists: "Freeways heavily hit by earthquake damage," UPI, *Lompoc Record*, February 10, 1971. Twenty-three years later, the San Diego Freeway did suffer structural damage from the 1994 Northridge Quake, primarily at its interchange with the Simi Valley Freeway, where connector lanes to the 118 were temporarily closed: "Drivers Beware," *Los Angeles Times*, January 28, 1994.

Accounts of the Lower Van Norman Dam are mostly culled from: the April/May 1971 issue of *California Geology*; "Six Interesting Facts About the 1971 Sylmar Earthquake," by Jason Rosenthal, the *Southern Californian* blog, February 9, 2015; and my interviews with Heckeroth. The reservoir was never refilled, mostly because construction of Castaic Lake, farther up the

5, eliminated the need, though that didn't stop some officials from trying to bring the Van Norman back online. Referencing the near-catastrophe of the dam, County Supervisor Ernest E. Debs asserted that such a proposal "just seems stupid": "Supervisor Calls Plan to Rebuild Dam 'Stupid,'" by Charles R. Donaldson, *Los Angeles Times*, December 22, 1971.

In the sidebar, the sniper along the 101 Freeway is mentioned in World Heritage Encyclopedia, among others. My source for the incident partially inspiring *Targets* is Peter Bogdanovich himself, gleaned from my days working for the director in 1991–1992. Several cinema blogs also relate this link.

Traffic on L.A. freeways has been a long-standing gag on *SNL* and late-night comedy shows. David Letterman's San Diego Freeway reference is recapped in "Top 10 David Letterman Late Show San Diego, Tijuana Jokes," by Matthew T. Hall, *San Diego Union-Tribune*, May 20, 2015.

O.J.'s slow-speed chase was a national phenomenon with hundreds of media sources. I largely pull from "Simpson surrenders after bizarre chase across L.A.," by Shawn Hubler and Jim Newton, June 18, 1994, and "The Sequence of Events," June 18, 1994, both from the *Los Angeles Times*. Also "Wild chase glues nation to TV sets," by Lynn Elber, AP, *San Bernardino Sun*, June 19, 1994, and portions of the CNN broadcast from June 17, 1994. The TV audience figure ("95 Million Watched the Chase") is sourced from AP, *New York Times*, June 22, 1994. The banter between Hal Fishman and the Skycam 5 pilot are taken from KTLA's live broadcast on June 17, 1994, available online.

Sidebar: So many Latino migrants were killed by vehicles in the twenty miles of freeway north of Oceanside, the corridor earned the terrible nickname "Slaughter Alley." Often they were dropped off by smugglers south of the checkpoint, then told to cross the lanes and walk through Camp Pendleton—getting struck before they could rejoin smugglers on the northern side—or they simply panicked upon seeing the patrol station and bailed out in the middle of the freeway ("Immigration: Slaughter Alley," by Raoul Contreras, *Reason*, June 1991). The *Los Angeles Times* is among those that casually invoked the term "Slaughter Alley" ("Roadblocks Slow Freeway System," by Ray Hebert, *Los Angeles Times*, January 3, 1967). As for the giant yellow "Caution" signs: "With only one left, iconic yellow road sign showing running immigrants now borders on the extinct," by Cindy Carcamo, *Los Angeles Times*, July 7, 2017.

Accounts of the cursed San Diego-Ventura Freeway Interchange: "3 Interchange Projects Won't End the Gridlock," by Annette Kondo, *Los Angeles Times*, May 20, 2001, and "$50 Million in Band-Aid Fixes Seen for L.A. Traffic Quagmire," by Annette Kondo, *Los Angeles Times*, May 20, 2001. The latter also features the Getty Center ad.

The carpool controversy is covered in the Orange County edition of the *Los Angeles Times* over the first few months of 1990. Several sources point to

SR-73 as the first publicly operated SoCal toll road, including "$50 Million in Band-Aid Fixes Seen for L.A. Traffic Quagmire."

The closures of the San Diego Freeway through the Sepulveda Pass are largely sourced from MTA's website. Funding figures are partly cited from "Davis Unveils New Addition to Area's Carpool Lane Network," by Caitlin Liu, *Los Angeles Times*, February 23, 2002.

Much of the section on Carmageddon, Carmageddon II, and Jamzilla is informed by personal experience. Additional material comes from: "Coming soon: 'Jamzilla,'" by Martha Groves, *Los Angeles Times*, February 13, 2014; "Drivers mostly avoid 'Jamzilla' and I-405 work," AP, *Visalia Times-Delta*, February 18, 2014; "Detailed explanation of how the Mulholland Bridge over 405 will be torn down," by Steve Hymon, *The Source*, Metro.com, July 12, 2011; "Will Carmageddon sequel be worse?" by Ari Bloomekatz, *Los Angeles Times*, July 20, 2012; "L.A. freeway closure set for this weekend," by John Rogers, AP, *Visalia Times-Delta*, September 24, 2012.

The U.S. Department of Transportation's Federal Highway Administration provided details and updates on the Sepulveda Pass improvements. Incidentally, even their own website officially referred to the project as "Carmageddon"!

There are variations on the freeway renovation's final price tag. My figure is from the *New York Times*, "Los Angeles Drivers on the 405 Ask: Was $1.6 Billion Worth It?," by Adam Nagourney, December 20, 2016. INRIX's traffic data is cited in the *NBC-LA News* story "Traffic on 405 Freeway Got Worse Since Expansion Project, Study Shows," by Angie Crouch, May 7, 2019.

The *Hollywood Reporter* has a thorough write-up about the friends who dined on the empty freeway (sidebar): "Carmageddon: Story Behind Photo of Trio Dining on 405," by *THR* staff, July 18, 2011. I also conversed with photographer Jesse Glucksman.

Journalist Elizabeth Lopatto experiences Musk's Boring Company test-run (sidebar) for *The Verge* in "I Took A Ride Through Elon Musk's New Tunnel in California," December 19, 2018.

Measure M milestones are from Metro.net.

The idea of a train along the freeway in 1976 is raised in "Would Public Use Transit? Studies Offer Clues," by Eleanor Hoover, *Los Angeles Times*, May 23, 1976.

Future ExpressLanes and congestion pricing are sourced from MTA, as well as "County considers toll lanes for part of 405 Freeway," by Laura J. Nelson, *Los Angeles Times*, December 6, 2019, and "Metro considers congestion pricing for DTLA, I-10 Freeway, Santa Monica Mountains," by Steven Sharp, *Urbanize Los Angeles*, February 10, 2021.

CHAPTER 8: THE GLENDALE FREEWAY

The September 1, 1955, announcement of the freeway's name is sourced from "L.A. Has New Freeway—But in Name Only," Associated Press, *San Bernardino Sun*.

The 2010 Census lists Glendale as the fourth most populous city in L.A. County.

The traffic engineer's observation about Southern Pacific trains, as well as opposition by city officials and Van de Kamp's Bakery, appear in "Deal Charged on Overpass for Freeway," *Los Angeles Times*, December 22, 1955. Former Councilman Tom LaBonge shared with me the enticing aromas generated by Van de Kamp's back in the day.

For progress on the freeway through Glendale, including 1962 meetings, I rely on *California Highways* and the *Los Angeles Times*. The displacement figure of 963 families is taken from "Hearing Scheduled on Freeway Routing," by George Garrigues, *Los Angeles Times*, June 10, 1962. Glendale Mayor William Peters blasts Arnold C. Palmer in "Glendale Freeway Protest Overidden," AP, *San Bernardino Daily Sun*, August 24, 1962.

Silver Lake's rich early cinema history is deeply explored by many film blogs and publications, including "Storage Firm Buys Old Mack Sennett Studios," by Denise Hamilton, *Los Angeles Times*, June 25, 1987.

In addition to *California Highways* archives, the bulk of the 1962–1963 details about the proposed southern extension to the Hollywood Freeway is sourced from the *Los Angeles Times*, including: "700 Attend Hearing on Routing of Freeway," January 13, 1962; "Silver Lake Residents Protest Freeway Route," December 20, 1962; "New Glendale Freeway Link Will Bear Costly Price Tag," by Ray Hebert, January 28, 1963; and "Route for Part of Glendale Freeway OKd," January 24, 1963.

In the sidebar, the January 31, 1962, edition of the *Los Angeles Times* ("Patient Who Gave the Doctor a Pain," by Matt Weinstock) reveals that homes were being moved to the set of *To Kill a Mockingbird*.

The tunnel idea under the San Gabriel Mountains is spelled out in "Tunnel or Freeway for Valley-to-Desert Road?," by Don Snyder, *Los Angeles Times*, April 28, 1968 (which includes Lanterman's quotes), and "Officials Endorse Proposal for La Cañada-Antelope Valley Tunnel," by Robert Diebold, *Los Angeles Times*, March 13, 1968.

Frank Lanterman Freeway (sidebar): "Solon reluctant about honor," United Press International, *San Pedro News-Pilot*, June 16, 1978.

About the bottleneck on Verdugo Road, Glendale Mayor Kenneth Stevens is quoted in "Engineer Will Close 31-Year Career With Completion of Glendale Freeway," by Don Snyder, *Los Angeles Times*, January 22, 1978.

No slums in Glendale!: "New Freeway Paving Way for Economic Expansion," by Don Snyder, *Los Angeles Times*, April 23, 1967. Alpha Terrace advertisement: *Los Angeles Times*, June 22, 1980.

Accounts of the economic austerity slowing construction are sourced from: "Freeway Construction Survives Energy Crisis," by James Quinn, *Los Angeles Times*, March 10, 1974; "Freewheeling days a thing of the past," by Ray Hebert, *Los Angeles Times* (reprinted—*Honolulu Sunday Star-Bulletin & Advertiser*, March 25, 1973); and the Caltrans archives. Information about the *Isle of California* mural can be found at LAFineArtsSquad.com. Patt Morrison writes about John Meenan and his crew for the *Los Angeles Times* ("Crew Surmounts Opposition, Red Tape and Rocks to Build Freeway") in the May 30, 1976, edition.

The *Los Angeles Times* covers the final 4.4-mile link, including "Engineer Will Close 31-Year Career With Completion of Glendale Freeway," by Don Snyder, January 22, 1978.

The Caltrans official predicting the end of freeway-building is quoted in "Freeway Gaps Dramatize Money Crunch," *Los Angeles Times*, September 14, 1975.

The sidebar about freeway landscaping techniques is drawn from the *Los Angeles Times*: "Beauty, Ecology to Be Stressed in Glendale Freeway Landscaping," by Don Snyder, May 28, 1971, and "Faster Action Sought From State on Landscaping New Freeways," May 28, 1972. *California Highways* also discusses landscaping in pre-1967 issues.

The deletion of the 710 Freeway extension has been covered by many, including "710 Freeway Fighters Leave Lasting Legacy," by Ben Tansey, *South Pasadena News*, June 4, 2019.

The drama surrounding the threat of a southern extension is largely culled from my interviews with Heckeroth as well as the *Los Angeles Times*: "Frustration in Freeway Fight Grows," by Ray Hebert, August 21, 1972; "Freeway Construction Survives Energy Crisis," by James Quinn, March 10, 1974; and "Suburban Letters: Lack of Freeway Planning?," by Eleanor Wood (Information Officer, Caltrans), April 30, 1978. The profiles of Juan Vega and other neighbors are sourced from local dailies, including "Silver Lake vs. Caltrans: Neighborhood Fights for Its Homes," by Sam Kaplan, *Los Angeles Times*, January 22, 1979.

1980s accounts of renters buying back their homes are largely sourced from the *Los Angeles Times*: "Success Story: Renters Are Now Owners," by Sam Hall Kaplan, December 1, 1982, and "Tenants Get Chance to Buy Homes," by Alan Maltun, January 17, 1982. The co-op project is covered in "Subsidized Condos Aimed at Stabilizing Neighborhood," by Sam Hall Kaplan, *Los Angeles Times*, November 28, 1982.

The idea of reviving the extension (sidebar): "Freeway Bypass to Skirt Downtown Gaining Favor," by Larry Gordon, *Los Angeles Times*, May 29, 1986.

Improvement studies of the southern terminus have been outlined by Caltrans. Alternate traffic scenarios are discussed on various blogs and covered in: "Hipster Freeway? Echo Park's State Route 2 To See Controversial Improvements," by Dennis Romero, *LA Weekly*, August 29, 2012; "Re-imagining Glendale Boulevard," by Kristopher Fortin, *Streets Blog L.A.*, January 19, 2012; and "Proposal: Replace the 2 freeway with a transitway and housing," *Transiting L.A.*, March 28, 2018.

The mysterious case of dumped dirt (sidebar): "350 Tons of Evidence, Nary a Clue," by Amy Louise Kazmin, *Los Angeles Times*, September 26, 1991, and "Caltrans Invited Dumping Problem," letter by James Bonar, *Los Angeles Times*, February 14, 1991.

Larry Gordon profiled parents' concerns about the baseball field next to the freeway in "Community and Politicians Go to Bat for Ball Field," *Los Angeles Times*, March 26, 1987.

The KPCC poll was conducted on behalf of *Take Two*: "The Little Known History Behind LA's Most Tolerable Freeway," by Lori Galarreta, *LAist*, February 6, 2020. Brenda Rees spreads her Glendale Freeway gospel ("In Love With The 2: An ode to a Northeast L.A. freeway") on *Eastsider L.A.*, June 11, 2019.

As a post-script, a case can be made that the 134 West—the stretch just east of SR-2 where it cuts through the hills of Eagle Rock—may be true freeway nirvana. KCRW radio host Steve Chiotakis (@RadioChio) tweeted on January 30, 2021, that he "could see DTLA skyline to left, Century City skyline ahead, Pacific Ocean reflecting sun beyond that, and snow-capped mtns in rearview mirror." I would point out that one can also glimpse the Griffith Observatory, as others have added. A friendly debate (rare for Twitter) ensued in which Angelenos compared the virtues of both the 134 and 2.

CHAPTER 9: THE SANTA MONICA FREEWAY

The "Car-pool is bliss" message is included in "Imperfect 10," by Bill Boyarsky, *Los Angeles Times*, January 19, 1994.

My overview of the McClure Tunnel and Santa Monica is partly informed by *Santa Monica: A History on the Edge*, by Paula A. Scott (2004: Arcadia Publishing). A video archive of *Going Through the Tunnel*, by Thomas Edison, February 14, 1898, is preserved at the Library of Congress.

McClure's appointment in January 1954 is chronicled in "Knight Names Santa Monican To Highway Unit," *Sacramento Bee*, January 15, 1954.

Demolition and displacement figures are drawn by the aforementioned "Imperfect 10" article. As for the plight of Black families, the *Los Angeles Times* allowed "that a number of the families whose homes are threatened are of non-Caucasian ancestry and for this reason will encounter unusual difficulties in relocating" ("New Olympic Freeway Link Protested," March 24, 1954).

Pastor Stephen "Cue" Jn-Marie shared his perspective during our conversations and lived for a time in the Crenshaw district. As an exercise in contrast, the overwhelming sentiment of the early 1960s can best by summed up by the *Los Angeles Times'* Ray Hebert, the newspaper's renowned urban affairs writer: "In this land of diversity called California, probably nothing serves more as a great unifying force than the ribbons of concrete which criss-cross the state" ("Freeways Unifying Land of Diversity," November 26, 1962).

Sugar Hill's history is sourced mostly from "The thrill of Sugar Hill," by Hadley Meares, *Curbed LA*, February 22, 2018, and the *Los Angeles Sentinel*, which has published several entries about the neighborhood, including "The Sugar on the Hill . . ." by Brian W. Carter, August 8, 2012.

West L.A. opposition: "State Strong For Freeway Along Olympic Route," *Valley Times*, September 17, 1956; "Freeway Will Ruin Homes, Board Told," *Los Angeles Times*, September 30, 1955; and "Highway Division Firm in Freeway Route Stand," *Los Angeles Times*, September 15, 1956.

The division between Rees and McClure is profiled in "State Highway Official Raps Opposition," Associated Press, *Pomona Progress-Bulletin*, May 4, 1956.

The freeway's official name-change: "It's Santa Monica Freeway All the Way," AP, *San Bernardino Sun*, April 27, 1957.

Per 1960 Census data, St. Paul had 313,000 people. L.A.'s Westside had about 340,000.

In addition to *California Highways*, the report on federal funds for the freeway is derived from "Quick Action On Freeway Sought," *Los Angeles Times*, June 21, 1957. Details about right-of-way agents scoping out houses along the freeway corridor are from "500,000 Westside Freeway 'Orphans' to Join Frenzy," by Ray Hebert, *Los Angeles Times*, October 19, 1964 (also the source for the "Freeway Orphans" cited later).

Passages from the *Los Angeles Times* regarding the groundbreaking are taken from "West Side Arteries to Ease Motoring," by Norman Dash, April 14, 1957. Regarding Telford's quote, his colleague, Lyman R. Gillis, district engineer for the highway department, was a bit more prosaic: "What is really meant by [saying they're obsolete] is that the traffic volume desiring to use the new freeway facility is far in excess of the design capacity. People like the freeways too well" ("Traffic to Speed on Freeway Loop," *Long Beach Independent Press-Telegram*, November 3, 1959).

The first two openings in 1961–1962, per the *Los Angeles Times*: "Santa Monica Freeway Link Opened to Public," December 5, 1961, and "Downtown Loop of Freeway Opens," March 31, 1962.

In the sidebar, the pilfered asphalt roller is parlayed in "Police looking for missing asphalt roller," United Press International, *Redlands Daily Facts*, March 8, 1966.

The blimp was chronicled in "Santa Monica Freeway Link Put in Service," *Los Angeles Times*, January 30, 1965. I arrived at the 60,000 balloons calculation by extrapolating from this fun article: "Balloon Boy Helium Physics—How Much Helium Would It Take To Lift Falcon?," by Steve Spangler, Science.com, October 17, 2009. If my numbers are off, know that writers make notoriously bad mathematicians.

Benchmarks of the Santa Monica-San Diego Interchange are drawn from *California Highways* and "San Diego Freeway Snarls to Last a Year," *Los Angeles Times*, July 11, 1963.

Anne Bartolotti provided me with an excellent oral history of her mother, Marilyn Reece. Portraits of Reece are also based on: "Women in Transportation History: Marilyn J. Reece, Civic Engineer," TransportationHistory.org, March 15, 2019; "Marilyn J. Reece, 77: State's First Licensed Female Civil Engineer," by Dennis McLellan, *Los Angeles Times*, May 21, 2004; and "Freeway Builders Are Weekend Housewives," by Dorothy Townsend, *Los Angeles Times*, April 6, 1964. Kristen Stahl (sidebar) is profiled by Bartolotti.

To expand on the sign sidebar, the official sans-serif font on interstate signs is known as Federal Highway Administration (FHWA) Standard Alphabets for Traffic-Control Devices, or Highway Gothic for short. Versions rank from A to F, from narrowest to thickest.

The displacement of Black families in Santa Monica: "Negroes Getting Aid in Relocating in Santa Monica," *Los Angeles Times*, December 22, 1966. Several sources lay out the history of Santa Monica's bias against African Americans, including "How Racism Ruined Black Santa Monica," by Hadley Meares, *LAist*, December 23, 2020, and "In 'Whites Only' Era, an Oasis for L.A.'s Blacks," by Cecilia Rasmussen, *Los Angeles Times*, July 3, 2005. Pastor Jn-Marie also shared his insights on the matter. My characterization of Inkwell Beach (sidebar) is informed by several publications.

Santa Monica: A History on the Edge provided final tallies for the freeway, including the 119 traffic lights on Wilshire.

The award for Reece's interchange: "Freeway interchange wins award," *Redlands Daily Facts*, February 2, 1966.

The Diamond Lane Preview ads ran in March 1976 editions of the *Los Angeles Times*.

StrongTowns.org (by Rachel Quednau, October 8, 2015) lays out the carpooling campaigns during the war and features some great alarmist posters, including the one featuring the Hitler ghost.

A wealth of archives provide information about the Diamond Lane. Quotes and anecdotes through spring of 1976—including "Mad Monday," the day it opened—are mostly cited from: "Freeway Car Pool, Bus Lane Will Open Monday," by Ray Hebert, *Los Angeles Times*, March 14, 1976; "Diamond Lanes Stir Dissension in Caltrans," by Ray Hebert, *Los Angeles Times*, July 20, 1976; "Fast lane, 12-mile parking lot," AP, *Bakersfield Californian*, March 16, 1976; "Chaos On A Freeway," by John Kendall, *Los Angeles Times*, March 15, 1976; "'Fast' diamond lanes need bit more polish," AP, *Escondido Times-Advocate*, March 16, 1976; "Revolt on Diamond Lane," by Ray Hebert, *Los Angeles Times*, June 3, 1976; "Diamond Lane Painters in the Rough," by Jerry Cohen, *Los Angeles Times*, May 25, 1976; "Ward, Hahn Protest For Free—Follower Fined $5," *Los Angeles Times*, June 1, 1976; "Lane Shall Not Close by Congestion Alone," by Barry Siegel, *Los Angeles Times*, March 22, 1976; "Diamond freeway test smoother second day," *Redlands Daily Facts*, March 17, 1976; "Lawsuit Would Block Freeway Express Lanes," AP, *Sacramento Bee*, March 30, 1976; and "Diamond Lanes Blamed for Rise in Accidents on Nearby Streets," by Ray Hebert, *Los Angeles Times*, June 15, 1976. Harry Reasoner's bemused report for *ABC News* is from May 24, 1976.

By the summer of 1976, motorists had overwhelmingly rejected the Diamond Lane. In 2,300 letters to the *Los Angeles Times*, 89 percent were against it, with only 8 percent for it: "Diamond Lanes Stir Dissension at Caltrans," by Ray Hebert, July 20, 1978.

After a four-month study (March through June), City Traffic Engineer S. S. Taylor determined that street travel on adjacent streets to the freeway was up 13.1 percent to 17.2 percent during rush hours: "Diamond Lanes Stir Dissension at Caltrans," by Ray Hebert, *Los Angeles Times*, July 20, 1978.

Gianturco speaks frankly about her Caltrans experience in "Oral history interview with Adriana Gianturco," by California State Archives, Office of the Secretary of State, 1995. Gianturco backs the Diamond Lane program—and refutes damning data—in a controversial Op-Ed in the *Los Angeles Times*: "Diamond Lanes: No 'Plot' Against Public," June 24, 1976. Other Gianturco information is sourced from: "Adriana Gianturco fought a lonely battle for car-pool lanes in 1976. Now, they're part of the L.A. map," by Bettijane Levine, *Los Angeles Times*, February 22, 1994; "Adriana Gianturco," by Lucille Crespo, *Sacramento Bee*, October 4, 1976; and "Caltrans Chief Backs Alternatives in Transit," by John Morganthaler, AP, *Sacramento Bee*, November 9, 1976. As for her number-one political adversary, Councilman Zev Yaroslavsky defends his rudeness to Gianturco thusly: "With people as stubborn as she is, it doesn't pay to be diplomatic." "Caltrans Accused of Breaking Pact on Diamond Lane," by Ray Hebert, *Los Angeles Times*, October 1, 1976.

I also quoted Gianturco from a talk-show appearance she made on the PBS show *Voter's Pipeline*, which aired May 26, 1981.

The following sources document the court battles of the Diamond Lane: "Diamond Lane Goes Out Like a Lamb," by Ray Hebert, *Los Angeles Times*,

August 14, 1976; "Judge criticizes Caltrans on Diamond Lane attitude," AP, *Long Beach Independent Press-Telegram*, August 7, 1976 (although the judge ruled on August 10, his edict didn't take effect until August 13); "Santa Monica Freeway's Diamond Lane Suspended," AP, *Santa Cruz Sentinel*, August 11, 1976; and "Caltrans may fight diamond lane ruling," by Bill Packer, *Valley News*, October 20, 1976.

The death and aftermath of the Diamond Lane: "She's Still the Driver at Caltrans," by Ray Hebert, *Los Angeles Times*, March 4, 1977; "State, U.S. Officials Drop Fight Over Diamond Lane," by Ray Hebert, *Los Angeles Times*, January 15, 1977; "The Heat Is On Lackner Again," by the *Sacramento Bee's* Capitol Bureau, August 8, 1976; and aforementioned sources on Gianturco. Of note: Various Southern California representatives rejected Diamond Lane initiatives in their districts after the fiasco. Assemblyman William Campbell (R-Whittier) called it "a complete flop" and opposed its implementation on San Gabriel Valley freeways. "Campbell to Oppose Diamond Lanes," *Los Angeles Times*, July 29, 1976.

The Diamond Lane "sequel" on the San Diego Freeway (sidebar): "New Lane Eases Freeway Tieups," *Los Angeles Times*, February 2, 1977; "No Time Wasted in Using Diamond Lane," by Ray Hebert, *Los Angeles Times*, February 1, 1977; and "Gianturco Recounts Slings and Arrows," by Kim Murphy, *Los Angeles Times*, April 25, 1985. Adriana's name and happy face gracing the "Coming Soon" sign appear as a photo in a *Redlands Daily* editorial, February 5, 1977 (I also remember the odd message from my childhood).

Details of the Northridge Earthquake are from several sources. Among them: "Drivers Beware," *Los Angeles Times*, January 28, 1994; "L.A. Dodges Big One, But It's Bad Enough," by Tracey Kaplan and Marc Lacey (*L.A. Times* reporters), *Salt Lake Tribune*, January 18, 1994; and "I-10 Is Reopened—but Spans Need Retrofitting," by Virginia Ellis, *Los Angeles Times*, April 12, 1994. Data about traffic, economic cost, and construction costs are from "Santa Monica Freeway to Reopen on Tuesday," *Los Angeles Times*, April 6, 1994. The earthquake opportunists selling rubble are featured in "Only in L.A.," by Steve Harvey, *Los Angeles Times*, January 25, 1994, and "Santa Monica Freeway for sale . . . in pieces," *Santa Rosa Press Democrat*, January 29, 1994. The April 12 dedication reopening the freeway: "I-10 Drivers Celebrate Reopening of Freeway," by Nora Zamichow, *Los Angeles Times*, April 13, 1994.

The designation for Christopher Columbus is announced in "Southland Route Named for Columbus," *Los Angeles Times*, November 25, 1976. Mary McNamara profiles the Rosa Parks designation in "A Less Traveled Road to Fame," *Los Angeles Times*, July 10, 2002. The "undesignation" proposal is taken from Board of Supervisors public records.

The removal of statues has been well-covered in the news, including "Gov. Pete Wilson statue gone from San Diego's downtown after advocates call for removal," by Andrea Lopez-Villafana, *Los Angeles Times*, October 15, 2020.

CHAPTER 10: THE SIMI VALLEY FREEWAY

Two good resources for the early Simi Valley include "Santa Susana Pass State Historic Park Cultural Resources Inventory Historic Overview," prepared by Alexander D. Bevil, Historian, California State Parks, March–April 2007, and "The Chumash Era," by Ellen Darby, SimiHistory.com. "Simi"/"Shimiji" is explained in "In Simi Valley," by Beverly Moeller, *Valley News*, September 8, 1960.

The auto body paramedic is profiled in "Simi Valley Residents Now Getting Ambulance Service," *Oxnard Press-Courier*, March 13, 1957.

A parade of dailies bring Simi's early, rural years to life. Among them: "Simi Valley Has Big Tomato Crop," *Los Angeles Times*, October 23, 1929; "Simi Valley One of California's Most Productive Farm Areas," *Valley Times*, August 1, 1947; "Fig Festival Set for Today in Simi Valley," *Los Angeles Times*, June 12, 1927; "Earning While Learning," by E. Yale Waterman, *Valley Times*, June 22, 1930; "Simi Valley Will Select a Queen," *Ventura County Star-Free Press*, September 9, 1960; "Simi Valley Opposes Pen, *Ventura County Star*, February 5, 1930; and "Prison Row Stirs Simi," *Ventura County Star*, March 10, 1930.

The town's water issues are laid out in: "The Reader Speaks Out: Water," by R. E. Harrington, *Ventura County Star-Free Press*, June 18, 1956; "Simi Valley's Rich Future Depends on Water Supply," by Robert Mount, *Oxnard Press-Courier*, May 14, 1957; "Simi Valley Rationed on Water," Associated Press, *Los Angeles Times*, June 28, 1957; and "Simi Valley Eyes Annexation with L.A. for Water," *Ventura County Star-Free Press*, April 15, 1953.

The promotion of subdivisions throughout the '50s and '60s: "Simi Valley First Tract Now Open," *Van Nuys News*, October 2, 1955; San Simeon model homes advertisement, *Valley Times*, December 15, 1967; "Smog-Free Air Termed Simi Valley Inducement," *Valley Times*, November 18, 1955; and "In Simi Valley," by Beverly Moeller, *Valley News*, September 8, 1960.

"Blood Alley"—a phrase that shows up a lot in publications to describe dangerous roads in pre-freeway days—appears in "Henson To Urge State Road Building Speed Up," *Ventura County Star-Free Press*, January 1, 1964. Pro-freeway residents are portrayed in "Simi Freeway," by R. E. Harrington, *Ventura Star Free-Press*, November 10, 1960.

The bowling alley cocktail bar standoff is chronicled in the *Ventura County Star-Free Press*: "Minister Hits Proposed Simi Cocktail Lounge," October 31, 1961, and "Churches Vs. Liquor Debated in Simi Valley," November 1, 1961.

Mean old cuss: "Landowner Blocks County's Try At Simi Road Survey," *Ventura Star-Free Press*, August 19, 1964.

The 90 percent figure of commuters to L.A. (in 1964) gradually decreased over the years. By 1991, only 60 percent of motorists commuted to L.A. County for work. "Simi Leaders Protest State Freeway Delay," *Los Angeles Times*, September 25, 1966, and "Development Outpaces Road Improvements in 2 Counties," by Carlos V. Lozano, *Los Angeles Times*, April 22, 1991.

Riotous meetings about the freeway spilled onto the pages of the *Van Nuys News* ("Row Breaks Up Simi F'wy Meet: Jeers, Boos Wreck Discussion of Route," by Haig Keropian) on January 5, 1964, and the *Los Angeles Times* ("Simi Freeway") on November 29, 1964.

In addition to *California Highways*, the 1966 groundbreaking is covered in "Simi Valley Tracts Holding Open House," *Los Angeles Times*, April 24, 1966. Spacemaker model homes appear in *Times* ads that year; they are blurbed about in "A House That Can Grow Right Along With Family," April 16, 1967.

Blogs for Iverson Movie Ranch, Corriganville Movie Ranch, and Spahn Ranch (later Charles Manson's hangout) offer comprehensive accounts of the Westerns shot in Santa Susana Pass.

Kirst Construction: "Mountains Moved for Freeway," *Los Angeles Times*, August 29, 1966.

The July 1968 opening is covered in "Freeway to Give Simi Valley Growth Spurt," by Irv Burleigh, *Los Angeles Times*, July 7, 1968.

The complaints by Pacoima renters are captured in "Pacoima Community Plan Draws Fire at Meeting," by Richard Quist, *Van Nuys News*, January 7, 1973. Meanwhile, over in tonier Granada Hills, homeowners were resentful of paying higher property taxes—due to increased assessments—while being subjected to blight caused by the freeway. They formed a neighborhood group called SHAFT (State Highway Action for Taxpayers): "Homeowners Near New Freeway Fight Tax Hike," by Ken Lubas, *Los Angeles Times*, July 29, 1976.

The Ritchie Valens Memorial Highway dedication was covered by local journalists, including Dalia Espinosa of the *Los Angeles Daily News* on August 6, 2018.

The freeze on freeways and Wedworth's comments are from "2-Year Freeway Building Freeze Nips Planners," *Los Angeles Times*, March 4, 1973. I mostly access the *Times* for the freeway's political-football gamesmanship throughout the 1970s, including: "Reagan Kills Speedup of Simi Freeway Work," by Martha Willman, December 22, 1971; "Simi Freeway Petitioners Seek Meeting With Brown," by Ken Lubas, December 28, 1975; "Cline Rips Brown for Veto of Simi Freeway Speedup," October 2, 1975; "Simi Council Backs Proposed Two-Year, Two-Cent Gas Tax to Complete Freeway," by Ken

Lubas, December 18, 1975; "Caltrans to Seek Simi Freeway Funds Today," by Ken Lubas, July 15, 1976; and "Desired Changes Could Delay Simi Freeway Construction," August 29, 1974. Also sourced: "Brown Approves Three Highway Projects," AP, *Petaluma Argus-Courier*, January 7, 1977.

Noise complaints and sound walls (including in the sidebar) are primarily sourced from "Walling in noise on freeway," by Penelope Simison, *Valley News*, July 17, 1977.

The last link: "Last Simi Valley Freeway Link Will Be Dedicated Saturday," by Mark A. Stein, *Los Angeles Times*, December 16, 1982. The still-missing link: "Caltrans Calls for New Freeway Link," by James Quinn, *Los Angeles Times*, March 17, 1983.

Information about the Reagan Library and Museum is culled from the Ronald Reagan Presidential Foundation and Institute and "Reagan Library, Museum Ready to Take Center Stage," by Kenneth R. Weiss, *Los Angeles Times*, November 3, 1991. The library's effect on the community is tracked in "Little Trickle-Down Effect From Reagan Library," by Phil Sneiderman, *Los Angeles Times*, June 7, 1992.

Simi Valley's populace was 78 percent white during the 1992 trial; Blacks accounted for 1.5 percent ("The Rodney King fiasco town," by Carl Irving, *San Francisco Examiner*, May 1, 1992). Simi backlash—and the cops-per-capita stat—drawn from the same *Examiner* article, as well as "Simi Valley residents try to live down expectation," by Paul Pringle, Copley News Service, April 4, 1993.

The earthquake-damaged section of the freeway reopened on September 3, 1994—five months after the overpasses of the more esteemed Santa Monica Freeway were fixed. "Simi Valley Freeway Up, Running," by Jill Leovy, *Los Angeles Times*, September 4, 1993. "Simi Valley Freeway Damage," *Los Angeles Times*, January 21, 1994, also recaps damage.

Name-change to Ronald Reagan, from the *Los Angeles Times*: "Senate to Vote on Naming Freeway for Reagan," by Cynthia H. Craft, December 3, 1994; "118 Freeway May Honor Reagan," December 3, 1994; and "Signs Announce New Name for Freeway," by Jeff McDonald, August 1, 1995.

In the sidebar, the 118's first name-change can be officially fixed to August 1970: "Simi Valley Renamed by State," *Los Angeles Times*, August 16, 1970.

Simi Valley regularly comes up on "safest cities" rankings by the FBI and other sources. An *NBC News* piece from May 11, 2017—citing a study by the University of Vermont—deemed it the "fifth happiest" city in the U.S. (minimum 100,000 population).

CHAPTER II: THE MARINA FREEWAY

Comprehensive backgrounds on the Ballona Wetlands include: "The History of Ballona Wetlands," Friends of Ballona Wetlands; "Ballona Creek Has a Storied Past, but It's the Polluted Present That's the Problem," by Ron Russell, *Los Angeles Times*, February 13, 1994; and "The Lost Wetlands of Los Angeles," by Nathan Masters, KCET.org, February 29, 2012.

Multiple sources inform the Howard Hughes sidebar, including "Hughes Aircraft Company," part of the Wende Museum's Collection on the Cold War in Culver City.

The SSBA faction is profiled in a series of *Los Angeles Times* pieces: "Merchants Hit Artery Extension," September 20, 1959; "Three Culver Groups to Protest Routing of Marina Freeway Link," September 27, 1959; and "Work Start on Marina Freeway Set," October 11, 1964.

In addition to *California Highways*, Slauson Freeway cost estimates are expressed in "State to Present Route for Slauson Freeway," by Burt Wuttken, *Los Angeles Times*, December 15, 1966.

Johnny Carson's running "cut off your Slauson" gag (sidebar) is imprinted in my memory but can also be accessed from *The Tonight Show* archives and at IMDb and www.quotes.net/mquote/930670.

The flood control project: "Flood Control Work To Start," *Los Angeles Times*, November 6, 1960.

Rights-of-way are detailed in "Relocation of Freeway Interchange Under Study," *Los Angeles Times*, February 24, 1963. Deputy Delmar's ordeal is featured in "Freeway to Block Entry to Garage," by Mal Terence, *Los Angeles Times*, September 26, 1963.

The sidebar about UCLA students' findings is drawn from "Archeologists Say Bison Once Roamed in L.A.," United Press International, *San Bernardino Sun*, April 29, 1965.

Judge Wapner and the Assemblyman McMillan case are mostly constructed from: "L.A. Assemblyman Indicted on Freeway Bribery Charge," by Ron Einstoss and Art Berman, *Los Angeles Times*, April 15, 1965; "L.A. Legislator Denies 'Block Freeway' Bribe," *San Francisco Examiner*, March 26, 1968; "McMillan Testifies at Bribe Trial," by Ron Einstoss and Art Berman, *Los Angeles Times*, July 28, 1965; and "Assemblyman Is Acquitted Of Bribery," Associated Press, *Daily Independent Journal*, July 29, 1965.

The *People's Court* sidebar is informed by, among others, IMDb and "Rusty Burrell, 76; Bailiff in Real Life and on 'The People's Court,'" by Dennis McLellan, *Los Angeles Times*, April 19, 2002.

The portrait of Marina del Rey is drawn from various sources, including: "County to Dedicate Its $36.2 Million Marina," *Los Angeles Times*, April 4, 1965; "History of the Marina," by Frank Coffey, VisitMarinadelRey.com; "Marina del Rey Harbor Dedicated by Anderson," *Los Angeles Times*, April 11,

1965; and "County to Dedicate Its $36.2 Million Marina," *Los Angeles Times*, April 4, 1965. The 2006 proposal to bridge Lincoln: "Plan Puts Marina in Fast Lane," by Martha Groves, *Los Angeles Times*, August 23, 2006.

Besides *California Highways*, multiple sources mention the Pacific Coast Freeway (sidebar), all with varying information about the amorphous project's length.

A preview of the 1968 dedication: "Copter Will Open 1-1/2-Mile Freeway," *Los Angeles Times*, November 21, 1968.

The nugget about Nixon digging freeways is from "Nixon Cruises Freeways," AP, *Santa Cruz Sentinel*, August 27, 1973.

While a number of sources cover the freeway's name changes, I largely rely on: "Resolution to Name Nixon Freeway Killed," *Los Angeles Times*, July 4, 1969; "Assembly OK's Nixon Freeway," UPI, *Santa Rosa Press Democrat*, April 25, 1971; "Nixon Freeway to stay," UPI, *Argus*, August 27, 1974; "State Senate Votes to Rename Nixon Freeway," AP, *Los Angeles Times*, January 29, 1976; and "Nixon Freeway Exists No Longer," AP, *Long Beach Press-Telegram*, March 16, 1976. Arturo Salazar also discussed this issue with me. Readers' letters are from the December 26, 1972, issue of the *Los Angeles Times*.

Regarding the Nixon Freeway in Yorba Linda (sidebar), Michael Ballard from Southern California Regional Rocks and Roads clued me in on the freeway's confusing labels. Per California Streets and Highways codes, SR-90 in this stretch is technically Imperial Highway. The state ceded local control, so the city of Yorba Linda "owns" this portion. Since the city likes to promote its Nixon heritage, it named this section of highway after the ex-president. But that still doesn't explain why one side of traffic gets a "Richard Nixon Freeway" sign while opposing traffic gets "Richard Nixon Parkway"!

Interestingly, as "The Father of California Freeways," Senator Randolph Collier was against changing the Marina Freeway to the Richard Nixon Freeway, but not for reasons you might think. He held that naming freeways for people was confusing to out-of-towners. "I like numbers," he declared ("Name Hardest on Nixon Freeway," by Bill Stall, *Oakland Tribune*, September 28, 1971).

The "sailboat" dedication: "Final Link Tying Marina-Freeway to Be Dedicated Today," by Skip Ferderber, *Los Angeles Times*, March 30, 1972.

Kenneth Hahn's crusade to fix the Culver Boulevard intersection is detailed in: "Death Rides 'Freeway to Nowhere,'" by Ray Hebert, *Los Angeles Times*, December 10, 1972; "Asks Closure," AP, *Sacramento Bee*, December 14, 1972; "Courts Get Plan to Speed Up Traffic Cases," by Charles R. Donaldson, *Los Angeles Times*, June 18, 1972; and "Freeway Junction Where Four Died Will Get Major Changes," by Ray Hebert, *Los Angeles Times*, January 5, 1973. Always one for a hyperbolic quote, Hahn called the intersection "an engineering monstrosity . . . no other freeway of its kind is so poorly engineered"

("Traffic Fatality Sparks Probe of Marina Freeway," *Los Angeles Times*, May 18, 1972).

The idea of a "Ballona Freeway" is covered in "Changing name of Marina Freeway to 'Ballona Freeway' is proposed," by Martha Groves, *Los Angeles Times*, September 3, 2013.

Yelp reviews courtesy of www.yelp.com/biz/marina-freeway-los-angeles. Heather Sundell swoons about the Marina Freeway in "L.A. Freeways Ranked from Mildly Soul-Sucking to Totally Unbearable," *Los Angeles* magazine website, July 25, 2016.

CHAPTER 12: THE CENTURY FREEWAY

Several portions of this chapter are gathered from my interviews with Caltrans's Heinz Heckeroth, who was assigned oversight of the freeway in 1967. The rise and fall of Cold War-related jobs throughout this chapter—including employment figures—are culled from a variety of Southland newspapers from the early 1990s and Britannica Online.

Hawthorne City Manager Kenneth Jue is often credited with the "100 years to build" quip, though it appeared in multiple outlets, including: "L.A.'s Oft-Stalled Century: The Last of The Freeways," by William Trombley, *Los Angeles Times*, August 31, 1981; "Country's most expensive freeway opens in L.A.," *Sacramento Bee*, October 15, 1993; and "Festivities open Century Freeway work," by Eric Bailey, *Daily Breeze*, May 2, 1982 (in which County Supervisor Deane Dana gets a similar jab in).

My calculations for how long it took to build the freeway compared to other iconic projects are informed by various encyclopedic sources, as well as "Officials say 35-year struggle will pay off," by Jim Radcliffe, *Daily Breeze*, October 10, 1993.

"World's most litigated highway" appears in numerous publications, including "Snarled in Disputes for Years, Century Freeway Beginning to Take Shape," by Ray Hebert, *Los Angeles Times*, January 23, 1986.

Collier and Backstrand make their pitch for the freeway in "New Freeway Patroling Setup Urged," *Los Angeles Times*, November 8, 1955.

The Los Angeles Conservancy provided background on TRW's Space Park complex.

The name's genesis is mentioned in "Officials say 35-year struggle will pay off," by Jim Radcliffe, *Daily Breeze*, October 10, 1993.

Debates about the freeway's route in 1963 are articulated in: "South Route Adopted for Century Freeway," by Jerry Gillam, *Los Angeles Times*, November 18, 1965; "Opposition Mounts to Freeway Plans," by John R. Murphy, *Los Angeles Times*, June 2, 1963; and "Century Freeway Route Held Peril to Watts Towers," *Valley Times Today*, June 6, 1963. The Pacific Coast

Freeway proposal (sidebar): "Opposition Mounts to Freeway Plans," by John R. Murphy, *Los Angeles Times*, June 2, 1963.

In addition to Heckeroth's retelling, various *San Francisco Examiner* entries speak of the deletion of San Francisco freeways, as well as "Removing Freeways—Restoring Cities: San Francisco, CA," by the Preservation Institute. The transference of funds to the Century Freeway is reported in "State Will Seek To Save Federal Freeway Fund," *Sacramento Bee*, March 24, 1966, and "U.S. Funds Indicated for Century Freeway," *Los Angeles Times*, December 18, 1967.

Portraits of the corridor cities are thoroughly investigated by local papers: "Norwalk, Downey Split Sharply Over Century Freeway Routing," by Ralph McClurg, *Long Beach Independent Press-Telegram*, October 30, 1966; "1,100 Attend Century Freeway Hearing," by Ralph McClurg, *The Independent*, March 31, 1967; "Century Freeway Path Proposal Draws Protest," by Ray Hebert, *Los Angeles Times*, April 17, 1968; and "Two Cities Involved Oppose Elevated Century Freeway," by William J. McCance, *Los Angeles Times*, August 31, 1969.

Wedworth's defiance is drawn from "Freeway 'to Please Everybody' Doesn't," by Ray Hebert, *Los Angeles Times*, November 15, 1970. Wedworth's battle with Inglewood is drawn from "Two-City Parley Requested on Dispute Over Century Freeway," *Los Angeles Times*, October 25, 1970.

The Del Aire drama got a lot of local play, recapped in "Last of L.A. freeways opens," Associated Press, *News Journal*, October 15, 1993, and "Freeway 'to Please Everybody' Doesn't," by Ray Hebert, *Los Angeles Times*, November 15, 1970. The Keiths are featured in "Foe won't be cheering on Thursday," by Jim Radcliffe, *Daily Breeze*, October 10, 1993.

As an aside, state right-of-way agents admitted that some families had to move several times due to rerouting of the freeway. One family moved when their house was condemned for LAX. Two years later, they had to uproot again because of the freeway: "Freeways—'Octopus' Still Growing," by Ray Hebert, *Los Angeles Times*, June 28, 1971.

The "'Freedom Fighters'" plight—and Judge Pregerson's involvement—are covered in, among others: "Adamant Woman Triggers Halt to Century Freeway," *Los Angeles Times*, March 11, 1973; "Court orders hearings on Century Freeway," United Press International, *Redlands Daily Facts*, January 27, 1975; "New Century Freeway Hearing Upset by Court," by Ray Hebert, *Los Angeles Times*, December 6, 1973; "L.A.'s Oft-Stalled Century: The Last of The Freeways," by William Trombley, *Los Angeles Times*, August 31, 1981; and "The Last Freeway," by Hillel Aron, *Slake*, July 2011.

Attorney John R. Phillips echoing Pregerson's findings: "Century Freeway to Airport OKd by U.S.," by Ray Hebert, *Los Angeles Times*, October 18, 1978, and the aforementioned "L.A.'s Oft-Stalled Century: The Last of The

Freeways."

The 1972 EIR report is cited by various, including "Order Halting Century Freeway Brings Chaos," by Ray Hebert, *Los Angeles Times*, July 17, 1972.

Several reporters of the era investigate the ill effects of shutting down the freeway and resultant crime, including Kim Kowsky ("Vice Thrives Along Decaying Imperial Highway," *Los Angeles Times*, May 6, 1990) and Mike Jelf ("Century Freeway work halted by legal tangles," *Long Beach Independent Press-Telegram*, November 19, 1972).

Lynwood's riches-to-rags tale is retraced mostly via: "Century Freeway to Airport OKd by U.S.," by Ray Hebert, *Los Angeles Times*, October 18, 1978; "Hi-tech freeway opening in L.A., finally," by Bob Ivry, *San Francisco Examiner*, October 12, 1993; and "Fleas Blamed on House Moving by Caltrans," by Lee Harris, *Los Angeles Times*, August 22, 1982.

Caltrans's alleged threat to ram the freeway through Hawthorne on pillars is from "Officials say 35-year struggle will pay off," by Jim Radcliffe, *Daily Breeze*, October 10, 1993 (also the source for the Hawthorne Bell). Meanwhile, Governor Jerry Brown was so fed up by the Century Freeway, he backed an idea of converting it to a "mini-freeway" of only four total lanes, restricted mostly to buses and three-person carpools: "Scaling down of Century Freeway plans proposed," AP, *San Bernardino Sun-Telegram*, December 23, 1975.

The terms of Pregerson's consent decree are sourced from public records and newspapers, including: the aforementioned "L.A.'s Oft-Stalled Century: The Last of The Freeways"; "Snarled in Disputes for Years, Century Freeway Beginning to Take Shape," by Ray Hebert, *Los Angeles Times*, January 23, 1986; "Objections to Housing Plan May Stall Work on Century Freeway," by Bill Boyarsky, *Los Angeles Times*, September 30, 1980; and "Welcome nation's most expensive freeway," by Sharon Ching, *San Bernardino Sun*, October 17, 1993.

Incidentally, as light rail was being discussed, Assemblyman Bruce Young (D-Cerritos) proposed a "high-speed cable car system" for L.A. County ("Lawmaker Vows to Get Freeway Built," by Kristina Lindgren, *Los Angeles Times*, December 28, 1980). And you thought Elon Musk's hyperloop was a nutty idea!

As for the employment mandate dictated by the consent decree: By 1985, 10,000 applicants registered with the freeway's Employment Center. Thanks to administrative snafus, fewer than 250 landed jobs ("Snarled in Disputers for Years, Century Freeway Beginning to Take Shape," by Ray Hebert, *Los Angeles Times*, June 23, 1986).

The 1982 groundbreaking is captured in "Festivities open Century Freeway work," by Eric Bailey, *Torrance Daily Breeze*, May 2, 1982.

Toxic waste-not whatnot: "Freeway Trivia," *Daily Breeze*, October 10,

1993; "Caltrans Set for Toxic Dump Cleanup to Clear Route for Century Freeway," by Ray Hebert, *Los Angeles Times*, December 22, 1986.

The *Los Angeles Times* explores the lack of traction by minority/women's businesses in "Minority Haulers Called Front for White Business," by Ronald B. Taylor, March 4, 1991, and "High Failure Rate Haunts Minority Contract Efforts," by William Trombley and Ray Hebert, December 30, 1987. Pregerson's quotes in the caption are from the latter.

Multiple sources, including Caltrans, outline the details about light rail (announced in the summer of '84).

Regarding the "first" five-level interchange: Technically, there was already a five-level built in L.A. County in 1975. The unsung Foothill/Ventura/Long Beach exchange in west Pasadena is interlaced with five levels, but because some of it is below-grade, it's not easy to perceive until you drive through it ("Completion of Freeway Link to Create Nonstop Tie Between Two Valleys," by Don Snyder, *Los Angeles Times*, January 5, 1975). See also "Letters to the Times: Construction of Century Freeway," Don Squires, *Los Angeles Times*, September 11, 1981.

Details of the Century-San Diego Interchange are varied, but I mostly source "Fun's just begun on SD Freeway," by Mark Igler, *Daily Breeze*, July 16, 1989, and "Smoothing Traffic Flow Tests Freeway Interchange Builders," by Tim Waters, *Los Angeles Times*, September 7, 1986.

The Century-Harbor Interchange is largely informed by "Soaring Interchange on Century Freeway to Be One of a Kind," by Ronald B. Taylor, *Los Angeles Times*, December 10, 1989. Another stunning factoid: the monster interchange is also 1.5 miles wide and long, and contains "nine miles of cloverleaf loops, transition lanes and connecting roads." Dead bodies at the MCM site: "Taggers Rule," by Faye Fiore, *Los Angeles Times*, December 6, 1990.

TIME has a nice write-up about *La La Land*'s shoot on the HOV ramp: "Here's How They Made *La La Land*'s Extravagant Opening Musical Number," by Raisa Bruner, December 9, 2016.

The October 1993 dedication is featured in "New Los Angeles freeway opens; traffic moves 'full speed ahead,'" *Los Angeles Daily News*, October 15, 1993, and "Country's most expensive freeway opens in L.A.," *Sacramento Bee*, October 15, 1993. Also, "Naming of Anderson called 'fitting tribute,'" *News-Pilot*, October 15, 1993. American Indian photo: "Incense Before Smog," by Kevork Djansezian, AP, *San Francisco Examiner*, October 15, 1993.

Anderson's bio in the sidebar: "Anderson: The Man Who Got Things Done: Obituary," by Ted Johnson, *Los Angeles Times*, December 15, 1994.

The follow-up with Ester Keith is drawn from "Foe won't be cheering on Thursday," by Jim Radcliffe, *Daily Breeze*, October 10, 1993, and "Last of L.A. freeways opens," AP, *News Journal*, October 15, 1993.

Speeding in the sidebar: "How 1994's 'Speed' Captured a Changing Los

Angeles," by Colin Marshall, KCET.org, November 11, 2016, and "The Bus in 'Speed' Wasn't Supposed to Land Like a 747," by Ian Failes, *Befores & Afters*, June 5, 2019.

Most Green (C) Line information is from the *Los Angeles Times* and the *News-Pilot*, years 1993 to 1995. Original data for the Green Line is taken from "$254.5-Million Rail Line OKd for Century Freeway," by William Trombley, *Los Angeles Times*, June 14, 1984. Future plans for the line: "The Green Line is 25 years old. Some thoughts on that," by Steve Hymon, *The Source*, Metro.com, August 12, 2020.

Parks, Playgrounds, and Beaches for the Los Angeles Region was drafted by Frederick Olmsted Jr., John Charles Olmsted, and Harland Bartholomew, 1930.

END FREEWAY

The housing situation in El Sereno is multi-sourced and was informed by my conversation with Pastor Stephen "Cue" Jn-Marie, as well as a column by Liam Dillon ("Activists wield bolt cutters in a tense L.A. neighborhood as poor families seize empty homes," *Los Angeles Times*, December 23, 2020).

Various news sources documented Eric Garner's death in 2014. I experienced the shutting down of the Hollywood Freeway in 2020 firsthand.

Caltrans put out their Equity Statement on December 10, 2020. The agency has publicly adopted plans to invest, repair, and engage in communities, with programs centered on walking, bicycling, and transit in yearly operating budgets.

PHOTO AND MAP CREDITS

Note: All uncredited photos and graphics were culled from the author's personal collection. All appropriate lengths were taken to secure proper photo credits and permissions. Any omissions or errors are deeply regretted and will be rectified upon reprint.

American Association of State Highway and Transportation Officials (AASHTO)
Page 273 (top)

Courtesy Richard Ankrom
Page 50

Automobile Club of Southern California Archives
Page 62

Courtesy Michael Ballard, Southern California Regional Rocks and Roads
Page 325 (top, bottom)

Copyright 1954 California Department of Transportation, all rights reserved, Page 38; Copyright 1957 California Department of Transportation, all rights reserved, Pages 129, 135; Copyright 1960 California Department of Transportation, all rights reserved, Page 123; Copyright 1961 California Department of Transportation, all rights reserved, Page 270; Copyright 1962 California Department of Transportation, all rights reserved, Pages 21, 81 (bottom), 188; Copyright 1963 California Department of Transportation, all rights reserved, Page 274; Copyright 1971 California Department of Transportation, all rights reserved, Pages 147 (top), 252; Copyright 1982 California Department of Transportation, all rights reserved, Page 287; Copyright (date unknown) California Department of Transportation, all rights reserved, Page 284

Courtesy of the California History Room, California State Library, Sacramento, California
Pages 88, 194 (left)

Courtesy of the California History Room, California State Library, Sacramento, California. Courtesy of donor Arnold Hylen
Page 68 (bottom left)

Creative Commons
Pages 25 (Metropolitan Transportation Engineering Board, CC BY 2.0), 53 (AmandaLeighPanda, Flickr, CC BY 2.0), 80 (Don Searls, Flickr, CC BY 2.0; Paul Haddad created graphical overlay), 86 (Channone Arif, Flickr, CC BY 2.0), 110 (Los Angeles Fire Department, CC BY-ND 2.0), 119 (Scott L, Flickr, CC BY-SA 2.0), 125 (Chris Yarzab, Flickr, CC BY 2.0), 151 (bottom) (A Syn, Flickr, CC BY-SA 2.0), 158 (bottom) (Person-with-No-Name, Flickr, CC BY 2.0), 162 (Don Searls, Flickr, CC BY 2.0), 169 (top) (Ron Reiring, Flickr, CC BY 2.0), 176 (Tobin, Flickr, CC BY-SA 2.0), 177 (biofriendly, Flickr, CC BY 2.0), 198 (Glenn Beltz, Flickr, CC BY 2.0), 204 (Arturo Sotillo, Flickr, CC BY-SA 2.0), 209 (Don Searls, Flickr, CC BY 2.0), 227 (haymarketrebel, Flickr, CC0 1.0), 229 (Marty B, Flickr, CC BY-SA 2.0), 239 (Erica Fischer, Flickr, CC BY 2.0), 260 (Steve Lyon, Flickr, CC BY-SA 2.0), 307 (NinaZafaz, Flickr, CC BY 2.0), 317 (Erica Fischer, Flickr, CC BY 2.0), 319 (joelorama, Flickr, CC BY-SA 2.0), 322 (Pedro Szekely, Flickr, CC BY-SA 2.0), 356 (Chris Yarzab, Flickr, CC BY 2.0)

Courtesy CSUDH Gerth Archives & Special Collections
Pages 109, 337 (top), 343

Dorothy Peyton Gray Transportation Library and Archive at the Los Angeles County Metropolitan Transportation Authority, CC BY-NC-SA 2.0
Pages 231 (Steve Hymon), 344, 353

Courtesy Ernest Marquez Collection. The Huntington Library, San Marino, California
Pages 60, 66

© 1993-2021 Jeff Gates
Page 346

Paul Haddad
Pages 15, 16 (left), 18 (top), 36 (left, right), 49, 81 (top), 89, 117, 124 (middle), 131, 138 (right), 142, 151 (middle left, right), 160 (left, right), 172, 185, 197, 205 (top), 233, 256, 257, 306, 310, 311, 327, 351, 352

Herald Examiner Collection/Los Angeles Public Library
Pages 67, 251 (bottom), 281

Landsat.com
Pages 132, 339 (bottom)

Danny Lloyd/Alamy Stock Photo
Page 354

Los Angeles Examiner Photographs Collection, 1920-1961/
USC Libraries Special Collections
Pages 99, 136

Courtesy Los Angeles World Airports
Page 350

Courtesy Novak Archive
Page 199

Public Domain
Pages 16 (right) (California Historical Society), 45 (Historical American Engineering Record, Library of Congress), 71 (California Historical Society), 78 (bottom) (U.S. National Archives and Records Administration), 149 (U.S. Geological Survey), 181, 222 (U.S. Geological Survey), 230 (U.S. Department of Transportation, Federal Highway Administration)

Public Domain—California Department of Public Works
Pages 23, 40, 43 (top, bottom), 44, 48, 63, 64, 68 (top, middle), 75, 77 (left), 78 (top), 97 (top, middle), 98, 102, 112, 116, 124 (bottom), 155 (middle), 156, 157, 164, 184, 186, 187, 193, 194 (top), 202, 210, 212, 214, 218, 241, 242, 243, 251 (top), 263, 268, 298, 300, 302 (top, bottom), 323, 332 (top)

Public Domain—University of Southern California Libraries and California Historical Society
Pages 6, 31 (bottom), 33, 39, 68 (bottom right), 94, 127, 128, 272

Courtesy San José State University Special Collections & Archives
Page 147 (bottom)

Copyright © Patrick Schnetzler
Page 232

Alexx Thompson
Page 408

Courtesy UCLA Library Special Collections, Charles E. Young Research Library
Pages 17, 20, 24, 26 (left), 35 (left, right), 41, 42, 61, 65, 72, 77 (right), 84, 96, 100, 103, 105, 107, 108, 113, 115, 124 (top), 130, 133, 138 (left), 145, 155 (top), 158 (top), 159, 163, 169 (bottom), 180, 195, 201, 213, 216, 219, 220, 235, 238, 240 (top), 240 (bottom), 247, 249, 254, 259, 261, 264 (top), 271, 273 (bottom), 275, 279, 280, 285, 293, 297, 305, 316, 324, 331, 332 (bottom), 335, 337 (bottom), 339 (top), 342, 355

Valley Times Collection/Los Angeles Public Library
Page 301

Wikimedia Commons
Pages 8 & 9 (Department of City Planning, PD), 18 (bottom) (PD), 26 (right) (PD), 27 (Bamsb900, CC BY-SA 4.0), 31 (top) (Pasadena Museum of History, PD), 51 (Steve Devorkin, PD), 56 (PD), 58 (PD), 73 (PD), 76 (PD), 92 (Dicklyon, CC BY-SA 4.0), 154 (PD), 93 (PD), 174 (PD), 175 (PD), 191 (Mike Dillon, CC BY-SA 3.0), 192 (Bobak Ha'Eri, CC BY-SA 3.0), 194 (right) (John M. DeMarco, CC BY 4.0), 205 (bottom) (Junkyardsparkle, CC0 1.0), 217 (Andreas Praefcke, CC BY 3.0), 264 (bottom) (PD), 267 (PD), 278 (PD), 289 (top, bottom) (FEMA, PD), 290 (The Jon B. Lovelace Collection of California Photographs in Carol M. Highsmith's America Project, Library of Congress, PD), 292 (Coolcaesar, CC BY-SA 3.0), 295 (Niceley, CC BY-SA 4.0), 304 (CC BY-SA 3.0), 309 (cropped) (Ricky Bonilla, CC BY-SA 2.0), 314 (PD), 338 (right) (HABS-Historic American Buildings Survey, PD), 341 (Remi Jouan, CC BY-SA 2.5), 345 (NGerda, PD), Back cover: (Bamsb900, CC BY-SA 4.0)

William Reagh Collection/Los Angeles Photographers Collection/
Los Angeles Public Library
Page 85

ACKNOWLEDGMENTS

I had been writing a book about L.A.'s freeways in my head since I was five years old without consciously realizing it. There are many parties who helped bring my obsession to life.

First and foremost, I want to extend my gratitude to Caltrans veteran Heinz Heckeroth. Getting to know Heinz over a series of phone calls and emails was like zipping along in a time capsule to the dawn of freeways. With nearly seventy years of transportation experience in Southern California, Heinz was involved as either an engineer or supervisor with almost every freeway in this book up to the Century Freeway. His anecdotes and vast knowledge breathed life into the spaces between the lines.

Thank you to Arturo Salazar, who is more than just a Caltrans Transportation Engineer. His addictive "Freeways of Los Angeles" on social media is a fountain of often obscure archives fashioned over decades. Arturo imparted his keen engineering insight and happily reached out to others on my behalf.

Special thanks to Karen Kasuba, Supervising Librarian II at the Caltrans Transportation Library, who guided me to Caltrans's digital trove of oral interviews and *California Highways and Public Works* back issues. Karen and her team helped source and approve Caltrans photos. Thanks also to Steve DeVorkin, Caltrans's resident TV Specialist, who helped me navigate the agency's media resources.

I particularly enjoyed my interview with Anne Bartolotti, daughter of Caltrans engineer Marilyn Jorgenson Reece. Anne's recollections truly painted a three-dimensional picture of her pioneering mother.

I am indebted to Pastor Stephen "Cue" Jn-Marie of The Row Church. Cue offered a personal perspective that I couldn't get from books to aid my understanding of how freeways disrupted underserved communities, and the multi-generation repercussions caused by their encroachment on these communities.

I would also like to acknowledge Richard Ankrom, Tom Philo (Librarian, CSUDH University Library), Michael Dolgushkin (Librarian, California State Library), and Craig Simpson (Director, Special Collections and Archives, SJSU) for their assistance and generosity. As the Corporate Archivist at the Automobile Club of Southern California, Morgan Yates shared access to materials that contextualized the region's transportation history. Thanks also to the Los Angeles Public Library's Photo Friends.

A special salute to Michael Ballard, who doesn't just run SoCalRegion.com, but offered his last-minute services as a "foot soldier" to snap the Nixon roadway signs in Yorba Linda—the final photos laid into this book!

My biggest appreciation goes out to my family, who provide the lifeblood for my creative pursuits.

ABOUT THE AUTHOR

PAUL HADDAD's books include the *Los Angeles Times* bestseller, *10,000 Steps a Day in L.A.: 57 Walking Adventures*, and *High Fives, Pennant Drives, and Fernandomania: A Fan's History of the Los Angeles Dodgers' Glory Years, 1977-1981* (named one of the Best Baseball Books of 2012 by the *Daily News*). As a Hollywood-born native, he has written about Los Angeles for the *Los Angeles Times* and hosted a column on *Huffington Post* about L.A.'s forgotten history. He has authored three award-winning novels, including the L.A. Noir *Paradise Palms: Red Menace Mob*. A graduate of University of Southern California's School of Cinematic Arts, Haddad has been nominated for multiple Emmys as a documentary producer. PaulHaddadBooks.com @la_dorkout

Patt Morrison is a journalist, best-selling author, and radio-television personality based in Los Angeles and Southern California. Morrison has a share of two Pulitzer Prizes as a longtime *Los Angeles Times* writer and columnist. As a public television and radio broadcaster, she has won six Emmys and a dozen Golden Mike awards.